Terrorism, Hot Spots and Conflict-Related Issues

MAINTAINING HOMELAND SECURITY

TERRORISM, HOT SPOTS AND CONFLICT-RELATED ISSUES

National Infrastructure: Protecting. Funding and Rebuilding
Frederick H. Lupul
2009 ISBN 978-1-60692-369-6

Wasted Lessons of 9/11
U.S. House of Representatives
2009 ISBN 978-1-60692-640-6

Maintaining Homeland Security
Alan P. Proctor and Jason T. Waleford
2009 ISBN 978-1-60692-990-2

Terrorism, Hot Spots and Conflict-Related Issues

MAINTAINING HOMELAND SECURITY

**ALAN P. PROCTOR
AND JASON T. WALEFORD
EDITORS**

Nova Science Publishers, Inc.
New York

For permission to use material from this book please contact us:
Telephone 631-231-7269; Fax 631-231-8175
Web Site: http://www.novapublishers.com

LIBRARY OF CONGRESS CATALOGING-IN-PUBLICATION DATA
Available upon request.
ISBN: 978-1-60692-990-2

Published by Nova Science Publishers, Inc. ✝ New York

CONTENTS

PREFACE

The U.S. Department of Homeland Security (DHS) became the nation's 15th Cabinet department in January 2003. Twenty-two agencies with different missions, operations and cultures were consolidated and realigned into one organization with its primary mission being that of prevention, preparation for and response to catastrophes affecting our homeland, while facilitating legitimate travel and trade (Appendix C). DHS now employs over 208,000 individuals who are located in every state and many foreign countries making it the third largest cabinet agency.

In: Maintaining Homeland Security
Editors: A. P. Proctor et al., pp. 1-28
ISBN: 978-1-60692-990-2

Chapter 1

HOMELAND SECURITY ADVISORY COUNCIL, REPORT OF THE ADMINISTRATION TRANSITION TASK FORCE, JANUARY 2008*

Department of Homeland Security

BACKGROUND

The U.S. Department of Homeland Security (DHS) became our Nation's 15th Cabinet department in January 2003. Twenty-two agencies with different missions, operations and cultures were consolidated and realigned into one organization with its primary mission being that of prevention, preparation for and response to catastrophes affecting our homeland, while facilitating legitimate travel and trade (Appendix C). DHS now employs over 208,000 individuals who are located in every state and many foreign countries making it the third largest cabinet agency.

In early 2007, in response to White House Executive Order 13286 (Appendix D), the Department of Homeland Security initiated activities at five strategy levels (Appendix E) to ensure operational continuity of homeland security responsibilities as part of planning for the January 2009 transition to a new presidential administration. This will be the first time that DHS has been required to meet the challenges of a presidential transition. Due to the critical nature of its

---

* This is an edited, excerpted and augmented edition of a Department of Homeland Security publication.

mission (Prevent, Prepare, Respond, and Recover), it is important that DHS take action now to ensure a seamless and agile transition to new leadership and optimization of the new leadership's ability to assume operational control of the Department.

Secretary Chertoff asked his Homeland Security Advisory Council (HSAC) to establish an Administration Transition Task Force (ATTF) to provide recommendations and best practices for the Department and its various homeland security partners to execute an efficient and effective administration transition. The chapter will also be shared with each presidential nominee in order to help prepare the incoming administration. The main objective of the ATTF chapter is to advocate strategies and policies that will strengthen operational continuity of the full spectrum of national homeland security operations. Briefings, research and recent history (e.g., United States - September 11, 2001, Spain - March 11, 2004, United Kingdom - June 30, 2007, etc.) have provided the ATTF an appreciation of the potential vulnerabilities during transition periods. Not only are we aware that they exist, but our enemies are as well. The ATTF believes that the most vulnerable period is 30 days prior to, as recent history has shown, and through six months after the change in administrations.

We recommend that the present Administration and Congress quickly implement the recommendations in this chapter and that the chapter be immediately passed to all appropriate entities, especially the presidential nominees.

The ATTF has endeavored to make this chapter and its recommendations politically neutral and to incorporate the expertise and experiences (best practices) of federal, state, local and private sector transitions. The ATTF believes the chapter will continue to foster and build collaborative support and respect from Congress, federal agencies, and state, local, tribal and private sector authorities.

The ATTF believes that the incoming and outgoing administrations must work closely together during the administration transition. It is extremely clear that successful transitions require a shared commitment to ensuring a smooth transition of power. This is facilitated by a positive attitude and open mind in both incoming and outgoing administrations, combined with the willingness to respect and listen to each other's concerns and priorities. This same attitude must also characterize the behaviors of the senior career personnel who remain with the Department and will be counted on to ensure a smooth transition between administrations.

It is important that the American public become engaged in understanding the unique vulnerabilities posed by this transition period. This will require public education and media engagement during this critical period in our history. Before, during and after the transition, the public must learn about the choices faced by the nation, communities, families, and individuals. The public must become a partner with their government, sharing the burden. In addition, DHS should continue to engage the media as an ally in the timely dissemination of accurate and actionable information. DHS must work with multiple messengers, trusted within diverse communities, to effectively communicate this information.

The ATTF is encouraged by early DHS efforts but much important work remains to be done.

PROCESS

The ATTF met in person and by teleconference between September and December 2007. Task Force members shared their own knowledge and experience in leading change and managing transitions within their own public and private organizations. The Task Force members also engaged public and private sector subject matter experts involved with both successful and unsuccessful organizational transformations and transitions. The following recommendations incorporate the knowledge and expertise of ATTF members and subject matter experts.

RECOMMENDATIONS

The recommendations are divided into seven broad categories. The ATTF recognizes that several of its recommendations could be aligned into multiple categories. The seven categories include: Threat Awareness, Leadership, Congressional Oversight/Action, Policy, Operations, Succession and Training. There is no rank order of recommendations within each category. We believe all constitute national imperatives and must be expeditiously implemented.

THREAT AWARENESS

Outgoing DHS Leadership should:

- Work with media partners to educate and inform the public that a period of heightened threat is likely before, during and shortly after the Presidential election and transition period.
- Clarify the meaning of "heightened threat" during the transition period by notifying all homeland security partners of historical patterns.

  - Provide timely and reliable dissemination of any credible threat reports to all presidential and vice presidential nominees
  - Encourage issuance of one joint statement on heightened threat level from all presidential nominees

- Enlist non-partisan/bi-partisan/neutral third parties and use public service announcements to assist in informing the public of increased threat levels and the rationale behind them.[1]
- Develop contingency plans around the now common themes of Prevent, Prepare, Respond, and Recover.

LEADERSHIP

Outgoing DHS Leadership should:

- Provide the presidential nominees with identified best practices and lessons learned domestically and internationally from analysis of incidents during leadership transitions.

  - Engage past White House Office of Homeland Security and DHS officials and transition teams at all levels of government (federal, state, local) and the private sector.
  - Engage the expertise of other federal departments' transition efforts with particular emphasis on the efforts of National Security organizations (e.g., Defense, State and Justice Departments).

- Work with the presidential nominees, their senior staff, and the Senate, prior to the election, to establish an expedited process for handling appointments and confirmation to critical assignments (this goes far beyond the top three or four senior positions in the Department). Encourage, with incentives (i.e., bonuses), current appointees to overlap the new administration term until the transition process is complete and new appointees are in place.

  - Draft lists of potential candidates for appointed positions in early summer. o Identify ways to accelerate the processing and Senate confirmation of Presidential appointments.
  - Ensure an increase in OPM investigative and adjudicative manpower to quickly clear senior and second-tier appointees (i.e., down to a minimum of Deputy Assistant Secretary positions).
  - Perform updates rather than completely re-do the clearance history for people already holding clearances (at least for all but very top positions). o Develop a framework for engaging all presidential nominees to ensure consistency on how they should interact with DHS and vice versa.
  - Ensure department-wide reciprocity for suitability that would allow for quicker movement between components.

- Encourage all presidential nominees to identify members and organize Homeland Security advisory groups in preparation for the administration transition.

  - Offer time and expertise from DHS HSAC membership to all interested presidential nominees and the President-Elect.

- Encourage, and where possible, obtain the commitment of current political appointees to remain until at least the end of the current administration. (Note: this recommendation is also under Congressional Oversight/Action)
- Hold personal meetings for outgoing leadership (Secretary, Deputy Secretary, etc.) with incoming leadership.
- Build and maintain a comprehensive list of DHS alumni of both political and senior career personnel for reference purposes.

- o Provide each incoming appointee, at the time he or she is nominated, with a complete list of recent predecessors/equivalents and their contacts (i.e. email, telephone, postal address, etc.).

- Implement further recommendation number one of the HSAC's Culture Task Force Report -- "DHS Headquarters Must Further Define and Crystallize Its Role." (Appendix F)
- Prepare an outreach strategy to federal, state, local, tribal and private sector leaders to accelerate the new senior leadership teams' ability to implement phone calls, meetings, etc. as soon as they officially assume their positions.
- Generate cost-benefit reports on the more controversial line items in the budget so that decisions can be made either to protect or remove prior to and through the transition process.

Incoming DHS Leadership should:

- Nominate and seek Congressional approval of the new Secretary of Homeland Security as is done with the Secretary of State and Secretary of Defense on the first day of the new Administration. (Note: this recommendation is also under Congressional Oversight/Action)
- Meet with federal, state, local, tribal, private sector, and media partners to discuss transition details.
- Ensure the current career Deputy Under Secretary for Management remains in this position during the next administration. (Note: the ATTF commends the Department for quickly appointing a senior career individual to this position.)

Congressional Oversight/Action

- Act with the same sense of urgency in considering and expeditiously approving the new Administration's Secretary of Homeland Security as is done with the Secretary of State and Secretary of Defense. (Note: this recommendation is also under Leadership)
- Form a select bipartisan group from existing Senate oversight committees to expedite confirmation for all incoming DHS nominees for national security positions with the deadline being the start of the August 2009 recess. (Note: we are NOT asking Congress to form another Committee.)

- Continue to update the Transition Act of 1963 as amended to reflect post-9/11 realities. (Appendix G).
- Implement 9/11 Commission recommendation to reduce the number of Congressional oversight committees and subcommittees from its current unwieldy eighty-six.
- Pass a Fiscal Year 2009 budget for the Department of Homeland Security much sooner than the Fiscal Year 2008 budget was passed to avoid negative impacts on operations and training that can result from continuing resolutions. Congress should also review the Department's FY 2008 budget to ensure sufficient resources are available and allocated for transition activities. This must include pre-election and post-election transition crisis management exercises. Budget shortfalls should be supplemented where necessary.

  - Fund crisis exercises at adequate levels prior to the transition period.
  - Establish critical line items for the budget.

- Continue work to reduce (with outgoing DHS leadership) the number of presidential-appointed senior positions at DHS. (Note: this recommendation is also under Succession)
- Provide early briefings and interactions with DHS presidential nominees and appointees detailing Congressional expectations with respect to homeland security responsibilities.
- Interact with presidential nominees in a bipartisan manner because homeland security is a non-partisan undertaking.
- Encourage incoming appointees to serve as consultants to DHS during their confirmation process.
- Encourage incoming DHS leadership to continue employing current appointees until they are replaced. (Note: this recommendation is also under Leadership)
- Discourage any reorganization of the department prior to or during the transition period. (Note: this recommendation is also listed under Operations)
- Consider current political appointees with highly specialized and needed skills for appropriate career positions. (Note: this recommendation is also under Succession)

Policy

Outgoing DHS Leadership should:

- Continue to encourage all homeland security partners to support and participate in transition efforts.
- Continue to enhance and build consensus among all partners (federal, state, local, tribal, private sector, Congress, etc.) around policy issues that are a priority to the outgoing administration.
- Prioritize critical policies with measurable benchmarks that need to be addressed prior to the change in administration.

  - Provide the incoming administration detailed "End of Appointment"/Departure reports, including lessons-learned, organizational, operational and program successes/failures, and objective/non-partisan recommendations to move forward.
  - Engage and provide a process and templates by which federal, state, local, tribal and the private sector authorities may submit to incoming DHS officials their list of priorities and compilation of 'decisions made' and 'decisions needed.'

- Continue to support the active involvement of the Council for Excellence in Government and the National Academy of Public Administration to make recommendations at all levels of government and the private sector for transition efforts.

Operations

Outgoing DHS Leadership should:

- Continue to vigorously support the establishment of state fusion centers with both funding and personnel. Listen to their specific information requirements necessary to empower state and local collaboration during the possible heightened threat period at the time of transition and throughout the new administration.
- Offer operational briefings to presidential nominees and their staff. Develop executive summaries of important issues for the nominees to consider.

- Develop a clear and concise communications strategy for transition planning and increase coordination through media representatives.
- Discourage any reorganization of the department prior to or during the transition period. (Note: this recommendation is also listed in Congressional Oversight/Action)
- Take advantage of the period from January through November 2008 as an important time to establish and standardize processes and procedures in consultation with state, local, tribal and private sector authorities. Refrain from trying to implement hasty requirements the last few months of the Administration.

SUCCESSION

Outgoing DHS Leadership should:

- Continue to ensure all key positions[2] currently filled by appointees have back up senior level career personnel for operational continuity and a more fluid transition process (Appendix H)[3]. This should also be coordinated with the Department's succession planning efforts to make certain that all key leadership positions are currently filled.
- Support and implement a cadre of individuals fully focused on transition with the leadership designation of Deputy Chief of Staff for Transition (DCST). Provide the DCST with a task force composed of representatives from each component and staff office.
- Generate a priority list of briefing materials and ensure they are in a consistent format, clearly and concisely written, well organized, and professionally presented.

  - Identify a departmental topic specialist for each functional area and major program and any associated working group assigned to it.
  - Make certain that incoming senior managers have quick references – issue papers – for each topic to prevent information overload.
  - List all of the existing cross functional working groups and the initiatives or programs on which they are working.
  - Allow personnel to do their jobs, as opposed to being consumed with briefings, through use of secure automated or web-based tools.

- Compile a list of all presidential and homeland security directives and strategies and show how each align or not with the others.
- Continue to reduce the number of senior political appointees so that there is a more even mix of career and presidential appointed senior positions to maintain continuity and historical knowledge.[4] (Note: this recommendation is also under Congressional Oversight/Action)
- Consider current political appointees with highly specialized and needed skills for appropriate career positions. (Note: this recommendation is also under Congressional Oversight/Action)

Note: The National Academy of Public Administration is providing key recommendations in this area.

TRAINING

Outgoing DHS Leadership should:

- Organize table top exercises (based upon DHS's top ten scenarios) for new administration officials as early as possible and assure adequate funding, preparation, and delivery of same.

Note: The Council for Excellence in Government is providing key recommendations in this area.

APPENDIX A: ADMINISTRATION TRANSITION TASK FORCE MEMBERS

- **Glenda Hood** – (Chair) Former Florida Secretary of State and Mayor of Orlando, Florida; President and CEO, Hood Partners, LLC
- **Don Knabe** – (Vice Chair) Chairman, Los Angeles County, Board of Supervisors
- **John Magaw** – (Vice Chair) Former Under Secretary for Security, Department of Transportation
- **Karen Anderson** – Former Mayor, Minnetonka, Minnesota, National League of Cities

- **Norm Augustine** – Former Chairman and Chief Executive Officer, Lockheed Martin Corp.
- **Kathleen Bader** – Textron Board Member
- **Nelson Balido** – President & Chief Executive Officer, Balido & Associates,
- **Frank Cilluffo** – Associate Vice President for Homeland Security, George Washington University
- **Dan Corsentino** – Former Sheriff, Pueblo County, Colorado
- **James Dunlap** – Former State Senator, State of Oklahoma
- **Ellen Gordon** – Former Administrator, Iowa Emergency Management Division
- **Dr. Doug Huntt** – President and Chief Executive Officer, Burlington United Methodist Family Services, Inc.
- **Phil Keith** – Retired Chief, Knoxville, Tennessee Police Department
- **Herb Kelleher**, Executive Chairman, Southwest Airlines Co.
- **Brian Leary** – Vice President of Design/Development, Atlantic Station & Jacoby Development, LLC
- **Patrick McCrory** – Mayor, Charlotte, North Carolina
- **Peggy Merriss** – City Manager, Decatur, Georgia, Past President International City- County Management Association
- **Karen Miller** – Commissioner, Boone County Government Center, Columbia, Missouri, Past President National Association of Counties
- **Judith Mueller** – Director of Public Works, Charlottesville, Virginia
- **Greg Principato** – President, Airports Council International – North America
- **Dr. Paul Roth** – Executive Vice President for Health Sciences, University of New Mexico
- **Dr. Roxane Cohen Silver** – Professor of Psychology and Social Behavior, University of California at Irvine
- **Jim Douglas** - Governor of Vermont, and past President, Council of State Governments
- **Rick Stephens** – Senior Vice President, Human Resources & Administration, The Boeing Company
- **Dr. Lydia Thomas** – Former President and Chief Executive Officer, Noblis
- **Joe White** – Chief Executive Officer, American Red Cross St. Louis Chapter
- **Bill Whitmore** – Chief Executive Officer, AlliedBarton Security Services
- **Houston Williams** – Principal Owner, Raven Oaks Vineyards and Winery
- **Thomas Wyss** – State Senator, State of Indiana

- **Allan Zenowitz** – Retired General and former FEMA Senior Official
- **Ex-Officio: Judge William Webster** (HSAC Chair) – Partner, Milbank, Tweed, Hadley & McCloy, LLP
- **Ex-Officio: Dr. James Schlesinger** (HSAC Vice Chair) – Chairman, Board of Trustees, The MITRE Corporation

HSAC Administration Transition Task Force Directors

Mike Miron, Director, State and Local Officials Senior Advisory Committee
Candace Stoltz, Director, Private Sector Senior Advisory Committee

HSAC STAFF
Doug Hoelscher, Director, Homeland Security Advisory Committees
Michael Fullerton, Deputy Director, Homeland Security Advisory Committees
Chuck Adams, Director, Emergency Response Senior Advisory Committee
Jennifer Myers, Executive Assistant
Kezia Williams, Senior Conference Coordinator

Apendix B: Administration Transition Task Force – Subject Matter Experts

- **Christian Beckner,** Senate Homeland Security and Governmental Affairs Committee (Majority side)
- **Michael A. Beland,** House Homeland Security Committee (Majority side)
- **Robert C. Bonner**, Gibson, Dunn and Crutcher LLP, former Commissioner Customs and Border Protection
- **Katja Bullock**, Special Assistant to the President for Presidential Personnel
- **Michael Donley,** Director of Administration and Management, Department of Defense
- **Jamie S. Gorelick,** Wilmer Cutler Pickering Hale and Dorr LLP, former Deputy Attorney General and former 9/11 Commissioner
- **Janet Hale**, Deloitte & Touche, former Under Secretary Management, DHS
- **Clay Johnson,** Deputy Director for Management, OMB
- **Tony Lord,** First Secretary, Justice and Home Affairs, British Embassy

- **Theresa Manthripragada,** Senate Subcommittee on Oversight of Government Management, the Federal Workforce and the District of Columbia (Minority side)
- **Pat McGinnis,** President and CEO, Council for Excellence in Government
- **Michael J. Russell,** House Subcommittee on Management, Investigations, and Oversight (Minority side)
- **Michael Stroud,** House Homeland Security Committee (Majority side)

APPRNDIX C: DHS VISION, MISSION AND STRATEGIC GOALS

Vision

Preserving our freedoms, protecting America ... we secure our homeland.

Mission

We will lead the unified national effort to secure America. We will prevent and deter terrorist attacks and protect against and respond to threats and hazards to the nation. We will ensure safe and secure borders, welcome lawful immigrants and visitors, and promote the free-flow of commerce.

Strategic Goals

- **Awareness** — Identify and understand threats, assess vulnerabilities, determine potential impacts and disseminate timely information to our homeland security partners and the American public.
- **Prevention** — Detect, deter and mitigate threats to our homeland.
- **Protection** — Safeguard our people and their freedoms, critical infrastructure, property and the economy of our Nation from acts of terrorism, natural disasters, or other emergencies.
- **Response** — Lead, manage and coordinate the national response to acts of terrorism, natural disasters, or other emergencies.
- **Recovery** — Lead national, state, local and private sector efforts to restore services and rebuild communities after acts of terrorism, natural disasters, or other emergencies.

- **Service** — Serve the public effectively by facilitating lawful trade, travel and immigration.
- **Organizational Excellence** — Value our most important resource, our people. Create a culture that promotes a common identity, innovation, mutual respect, accountability and teamwork to achieve efficiencies, effectiveness, and operational synergies.

Appendix D: White House Executive Order 13286 Section 88 Order of Succession

For Immediate Release
Office of the Press Secretary
August 13, 2007

Executive Order: Amending the Order of Succession in the Department of Homeland Security

By the authority vested in me as President by the Constitution and the laws of the United States of America, including the Federal Vacancies Reform Act of 1998, 5 U.S.C. 3345, *et seq*., it is hereby ordered as follows:

Section 1. Section 88 of Executive Order 13286 of February 28, 2003 ("Amendment of Executive Orders, and Other Actions, in Connection With the Transfer of Certain Functions to the Secretary of Homeland Security"), is amended by striking the text of such section in its entirety and inserting the following in lieu thereof:

"Sec. 88. Order of Succession.

Subject to the provisions of subsection (b) of this section, the officers named in subsection (a) of this section, in the order listed, shall act as, and perform the functions and duties of the office of, the Secretary of Homeland Security (Secretary), if they are eligible to act as Secretary under the provisions of the Federal Vacancies Reform Act of 1998, 5 U.S.C. 3345 *et seq*. (Vacancies Act), during any period in which the Secretary has died, resigned, or otherwise become unable to perform the functions and duties of the office of Secretary.

(a) Order of Succession.

i Deputy Secretary of Homeland Security;
ii Under Secretary for National Protection and Programs;
iii Under Secretary for Management;
iv Assistant Secretary of Homeland Security (Policy);
v Under Secretary for Science and Technology;
vi General Counsel;
vii Assistant Secretary of Homeland Security (Transportation Security Administration);
viii Administrator of the Federal Emergency Management Agency;
ix Commissioner of U.S. Customs and Border Protection;
x Assistant Secretary of Homeland Security (U.S. Immigration and Customs Enforcement);
xi Director of U.S. Citizenship and Immigration Services;
xii Chief Financial Officer;
xiii Regional Administrator, Region V, Federal Emergency Management Agency;
xiv Regional Administrator, Region VI, Federal Emergency Management Agency;
xv Regional Administrator, Region VII, Federal Emergency Management Agency;
xvi Regional Administrator, Region IX, Federal Emergency Management Agency; and
xvii Regional Administrator, Region I, Federal Emergency Management Agency.

"(b) Exceptions.

i No individual who is serving in an office listed in subsection (a) in an acting capacity, by virtue of so serving, shall act as Secretary pursuant to this section.
ii Notwithstanding the provisions of this section, the President retains discretion, to the extent permitted by the Vacancies Act, to depart from this order in designating an acting Secretary."

Sec. 2. Executive Order 13362 of November 29, 2004 ("Designation of Additional Officers for the Department of Homeland Security Order of Succession"), is hereby revoked.

GEORGE W. BUSH THE WHITE HOUSE, August 13, 2007.

Appendix E: DHS Administration Transition Five Level Strategy

DHS Transition Planning

DHS is engaged in activities at five levels to ensure operational continuity of homeland security responsibilities in view of the January 2009 administration change.

ORDER OF SUCCESSION - White House Executive Order and DHS Management Directives provide Delegations of Authority and Orders of Succession for the Secretary and Component Leaders.

DHS SUCCESSION PLANNING - Analysis of DHS critical positions to identify succession risk. For each political appointee, identify interim acting career executive. (DHS Components led by USM-OCHCO; NAPA)

INTER-AGENCY COLLABORATION - Relationships, protocols, and interfaces among Departments of Defense, Justice, Commerce, State, Homeland Security and others. Develop homeland security curriculum for senior officials. (Dep Sec, USM– OCHCO; CEG)

BEST PRACTICES - Learning from state and local governments as well as the private sector. (Homeland Security Advisory Council – Administration Transition Task Force)

ADMINISTRATIVE TRANSITION GUIDANCE - Transition Guidance Handbooks will address priorities and challenges relevant to major programs and operational areas. (USM–OCAO)

Appendix F: HSAC's Culture Task Force Report (January 2007) -- "DHS Headquarters Must Further Define and Crystallize Its Role"

Recommendation 1: DHS Headquarters Must Further Define and Crystallize Its Role

The Department of Homeland Security was created to align and integrate a number of existing component organizations and to develop a number of new organizational capabilities in order to provide a cohesive, integrated and operationally efficient means of protecting the homeland. Like most new organizations that result from an organizational consolidation, one of the initial and key organizational challenges is to define the role and accountability of the headquarters organization so that the desired alignment and integration are achieved, while retaining the strong operational focus of the component organizations. The cycle time and effectiveness for the alignment and integration can be significantly improved when there is a clear definition of the role and mission of the headquarters organization. Such a definition provides a framework that ensures the headquarters can focus on its role, authority and accountability while at the same time empowering the DHS component organizations to focus on their roles. To date, while much substantive progress has been made, there does not appear to be final and full clarity with respect to these respective roles. The DHS leadership needs to ultimately define the role of headquarters so that the operational component organizations can focus on their operational strengths, while the headquarters provides the overall policy,

supports integrating processes where appropriate to leverage individual component strengths, and creates the organizational alignment necessary for overall DHS success. It is important that DHS headquarters not assume final operational responsibility for component missions but rather take responsibility for providing the effective vision, policies and resources to ensure the successful execution of all component missions. One CTF member viewed the DHS Headquarters optimum to be similar to the small sticker on most of our computer hard drives – "Intel inside" – and that DHS Headquarters' operational footprint should be that small in relationship to its component agencies. The following figure delineates key suggested elements for the DHS headquarters:

Proposed Role of DHS Headquarters

- Establish overall DHS strategy and annual operational and financial performance objectives
- Insure DHS performance against operational and financial objectives through oversight of DHS Component Commands
- Actively engage with DHS Component Commands in their strategies, investments, and leadership development
- Rely on DHS Component Commands for Day to Day Execution of DHS objectives
- Sponsor and Lead DHS values, ethics and compliance standards
- Sponsor initiatives that have DHS wide impact on performance
- Manage shared DHS services
- Lead and coordinate interface with Congress and other governmental agencies and organizations
-
-

DHS HQ Supports Component Command Leadership & Performance

The CTF suggests the following guidelines with respect to the role of DHS Headquarters in its relationship to its component organizations:

- Build Trust, by:

o Clearly defining the DHS mission to the American People.
o Publicly and internally establishing Homeland Security goals and performance metrics so all can objectively see and measure Homeland Security success. - Ensuring measurable performance against goals and metrics.
o Acting as the coordinated voice to Congress for the Department, proactively engaging Congress in establishing goals and measuring performance.
o Sponsoring and leading DHS values, ethics and compliance standards.

- Empower the components, by:

- Beginning each week with a Leadership Meeting with the components. - Aligning component goals with Homeland Security objectives.

- Integrating within component organizations the best functionality, practices and innovations of other components.
- Actively investing in the activities, people and strategy critical to the ability of component organizations to meet their goals.
- Sponsoring activities and initiatives that have enterprise-wide impact on performance.

- Empower employees, by:

  - Dropping the buzzword "Human Capital" and replacing it with "employees" or "members" of DHS.
  - Requiring members of the headquarters to visit and to listen and respond to employees and engage and support groups outside the headquarters. All DHS presentations (component, state, local, tribal and the private sector) outside the Headquarters should result in formal action items to be addressed, and the actions taken, or decisions with respect to them, provided to their proponents within 30 days.

- Be a Good Partner, by:

- Visiting regularly, and listening to, its Homeland Security Partners at the local levels where policies, plans, and requirements are frequently implemented and where security and disaster responses are most often executed.
- Administering grant programs in collaboration and partnership with its Homeland Security Partners.

APPENDIX G: PRESIDENTIAL TRANSITION ACT OF 1963

Legislative History (*Source: Senate Committee on Government Affairs)* The Presidential Transition Act of 1963 (Public Law 88-277, 3 U.S. C. 102 note) was enacted to provide for the orderly transfer of executive power in connection with the expiration of the term of office of a President and the inauguration of a new President. Since the time the Presidential Transition Act was passed, transitions have grown more complex and cumbersome, often leaving the new administration without the head start it needs to begin governing on inauguration day.

On November 1999, the House of Representatives, by voice vote, passed H.R.3 137, a bill "to amend the Presidential Transition Act of 1963 to provide for training of individuals a President-elect intends to nominate as department heads or appoint to key positions in the Executive Office of the President." After the bill was referred to the Governmental Affairs Committee, the Committee made changes to it based on recommendations from experts in the area of Presidential transitions. The Senate version, S. 2705, the Presidential Transition Act of 2000, was introduced on June 8, 2000. It was cosponsored by Senators Lieberman, Akaka, Collins, Durbin, Levin and Voinovich. It was ordered favorably reported by the Committee on June 14, 2000. The House of Representatives adopted the language of S. 2705, incorporated it into H.R. 4931, and passed the bill by unanimous consent on September 13, 2000. The Senate passed H.R. 4931 by unanimous consent on September 28, 2000.

H.R. 4931 was signed by the President on October 12, 2000. 118 STAT. 3856 PUBLIC LAW 108-458-DEC. 17, 2004 Subtitle F-Presidential Transition

SEC. 7601. PRESIDENTIAL TRANSITION.

SERVICES PROVIDED PRESIDENT-ELECT.-Section 3 of the Presidential Transition Act of 1963 (3 U.S.C. 102 note) is amended-

(1) by adding after subsection (a)(8)(A)(iv) the following: “(v) Activities under this paragraph shall include the preparation of a detailed classified, compartmented summary by the relevant outgoing executive branch officials of specific operational threats to national security; major military or covert operations: and pending decisions on possible uses of military force. This summary shall be provided to the President-elect as soon as possible after the date of the general elections held to determine the electors of President and Vice President under section 1 or 2 of title 3, United States Code.”;

(2) by redesignating subsection (f) as subsection (g); and

(3) by adding after subsection (e) the following:

“(f)(1) The President-elect should submit to the Federal Bureau of Investigation or other appropriate agency and then, upon taking effect and designation, to the agency designated by the President under section 115(b) of the National Intelligence Reform Act of 2004, the names of candidates for high level national security positions through the level of undersecretary of cabinet departments as soon as possible after the date of the general elections held to determine the electors of President and Vice President under section 1 or 2 of title 3, United States Code.

"(2) The responsible agency or agencies shall undertake and complete as expeditiously as possible the background investigations necessary to provide appropriate security clearances to the individuals who are candidates described under paragraph (1) before the date of the inauguration of the President-elect as President and the inauguration of the Vice-President-elect as Vice President.".

(b) Sense of the Senate Regarding Expedited Consideration of National Security Nominees.-It is the sense of the Senate that-

1 the President-elect should submit the nominations of candidates for high- level national security positions, through the level of undersecretary of cabinet departments, to the Senate by the date of the inauguration of the President-elect as President; and

2 for all such national security nominees received by the date of inauguration, the Senate committees to which these nominations are referred should, to the fullest extend possible, complete their consideration of these nominations, and, if such nominations are reported by the committees, the full Senate should vote to confirm or reject these nominations, within 30 days of their submission.

(c) Security Clearances for Transition Team Members.-

1 DEFINITION.-In this section, the term "major party" shall have the meaning given under section 9002(6) of the Internal Revenue Code of 1986.

2 IN GENERAL.-Each major party candidate for President may submit, before the date of the general election, requests for security clearances for prospective transition team members who will have a need for access to classified information to carry out their responsibilities as members of the President-elect's transition team.

3 COMPLETE DATE.-Necessary background investigations and eligibility determinations to permit appropriate prospective transition team members to have access to classified information shall be completed, to the fullest extent practicable, by the day after the date of the general election.

(d) EFFECTIVE DATE.-Notwithstanding section 351, this section and the amendments made by this section shall take effect on the date of enactment of this Act.

3 USC 102 note

APPENDIX H

Annex A: Order for Delegation of Authority by the Secretary of the Department of Homeland Security (Pursuant to Executive Order 13442 (August 13, 2007))

1 Deputy Secretary for Homeland Security
2 Under Secretary for National Protection and Programs
3 Under Secretary for Management
4 Assistant Secretary of Homeland Security (Policy)
5 Under Secretary for Science and Technology
6 General Counsel
7 Assistant Secretary of Homeland Security (Transportation Security Administration)
8 Administrator of the Federal Emergency Management Agency
9 Commissioner of U.S. Customs and Border Protection
10 Assistant Secretary of Homeland Security (U.S. Immigration and Customs Enforcement)
11 Director of U.S. Citizenship and Immigration Services
12 Chief Financial Officer
13 Regional Administrator, Region V, Federal Emergency Management Agency
14 Regional Administrator, Region VI, Federal Emergency Management Agency
15 Regional Administrator, Region VII, Federal Emergency Management Agency
16 Regional Administrator, Region IX, Federal Emergency Management Agency
17 Regional Administrator, Region I, Federal Emergency Management Agency

Annex B. DHS Succession Order and Order for Delegation

U.S. Coast Guard	
Commandant	S
Vice Commandant*	C
Chief of Staff	C
Commander, Pacific Area	C
Commander, Atlantic Area	C
Federal Emergency Management Agency	
Administrator	S
Deputy Administrator and Chief Operating Officer*	S
Deputy Administrator, National Preparedness	S
Associate Deputy Administrator	C
Director, Office of Policy & Planning Analysis	N
Region V Administrator	N
Region VI Administrator	N
Region VII Administrator	N
Region IX Administrator	C
Region I Administrator	N
U.S. Secret Service	
Director	C
Deputy Director	C
Assistant Director, Administration	C
Assistant Director, Protective Operations	C
Assistant Director, Investigations	C
Assistant Director, Protective Research	C
Assistant Director, Human Resources and Training	C
Assistant Director, Inspection	C
Assistant Director, Government and Public Affairs	C
U.S. Immigration and Customs Enforcement	
Assistant Secretary	S
Deputy Assistant Secretary, Operations*	C
Director, Office of Investigations	C
Director, Office of Detention & Removal Operations	C
Director, Office of the Principal Legal Advisor	N
Deputy Assistant Secretary, Management	C
U.S. Citizenship and Immigration Services	
Director	S
Deputy Director*	N
Associate Director, Domestic Operations	C
Associate Director, National Security & Records Verification	C
Chief Financial Officer	C
Director, New York District	C

(Continued).

U.S. Coast Guard	
U.S. Customs and Border Protection	
Commissioner	S
Deputy Commissioner*	C
Chief, Border Patrol	C
Assistant Commissioner, Field Operations	C
Director, Field Operations, New York	C
Sector Chief, El Paso	C
Director, Field Operations, Houston	C
Sector Chief, Tucson	C
Sector Chief, San Diego	C
Director, Field Operations, Miami	C
Transportation Security Administration	
Assistant Secretary / Administrator	S
Deputy Administrator*	C
Assistant Administrator, Office of Transportation and Sector	L
Assistant Administrator, Office of Security Operations	C
Assistant Administrator, Office of Law Enforcement/	C
Federal Security Director, Los Angeles International Airport	C
Federal Security Director, Orlando International Airport	C
Management	
Under Secretary	S
Deputy Under Secretary*	C
Chief Financial Officer	S
Chief Information Officer	P
Chief Human Capital Officer	P
Chief Procurement Officer	C
Chief Administrative Officer	C
Science and Technology	
Under Secretary	S
Deputy Under Secretary*	C
Director, Office of Transition	C
Director, Interagency Programs	L
Director, Office of Innovation	C
Division Head, Office of Explosives	C
Division Head, Office of Borders & Maritime Security	C
National Protection and Programs Directorate	
Under Secretary	S
Deputy Under Secretary*	P
Assistant Secretary, Infrastructure Protection	P
Assistant Secretary, Cybersecurity & Communications	N

U.S. Coast Guard	
Assistant Secretary, Intergovernmental Affairs	N
Deputy Assistant Secretary, Infrastructure Protection	N
Director, U.S. Visitor & Immigrant Status Indicator	C
Office of Policy	
Assistant Secretary	S
Deputy Assistant Secretary*	N
Assistant Secretary, Policy Development	N
Assistant Secretary, International Relations	L
Director, Screening Coordination Office	L
Office of Intelligence and Analysis	
Under Secretary, Chief Intelligence Officer	S
Deputy Under Secretary*	C
Deputy Under Secretary, Operations	N
Assistant Deputy Under Secretary, Intelligence	C
Assistant Deputy Under Secretary, Mission Integration	C
Assistant Deputy Under Secretary, External	C
Office of Operations Coordination	
Director	L
Deputy Director	C
Director, National Operations Center	C
Chief of Staff	C
Director, Incident Management & Interagency Planning	C
Office of Health Affairs	
Assistant Secretary, Chief Medical Officer	S
Principal Deputy Assistant Secretary, Deputy Chief Medical	C
Chief of Staff	C
Associate Chief Medical Officer, Medical Readiness	T
Associate Chief Medical Officer, Component Services	C
Associate Chief Medical Officer, Weapons of Mass	C
Federal Law Enforcement Training Center	
Director	C
Deputy Director	C
Office of the General Counsel	
Assistant Director, Training	C
Assistant Director, Field Training	C
Assistant Director, Training Innovation & Management	C
Assistant Director, Administration	C
Assistant Director, Chief Financial Officer	C
Assistant Director, Chief Information Officer	C
Senior Associate Director, Washington Operations	C
Domestic Nuclear Detection Office	
Director	P

(Continued).

U.S. Coast Guard	
Deputy Director	C
Assistant Director, Mission Management	C
Assistant Director, National Technical Nuclear Forensics	C
Assistant Director, Transformational & Applied Research	C
Assistant Director, Product Acquisition	C
Office of the General Counsel	
General Counsel	S
Principal Deputy General Counsel*	N
Deputy General Counsel	C
Associate General Counsel, General Law	N
Chief Counsel, TSA	C
Director of Field Legal Operations, Principal Legal Advisor,	C
Office of Legislative Affairs	
Assistant Secretary	N
Deputy Assistant Secretary, Operations	C
Deputy Assistant Secretary, Senate Liaison	N
Deputy Assistant Secretary, House Liaison	N
Director, Intelligence & Analysis/Operations	N
Director, National Protection & Programs Directorate	C
Office of Public Affairs	
Assistant Secretary	N
Deputy Assistant Secretary	N
Director, Strategic Communications	N
Director, Internal Communications	C
Office of Inspector General	
Inspector General	S
Deputy Inspector General*	C
Counsel to the Inspector General	C
Assistant Inspector General, Audits	C
Assistant Inspector General, Investigations	C
Assistant Inspector General, Inspections	C
Citizenship and Immigration Services Ombudsman	
Ombudsman	N
Executive Officer	C
Chief, Programs, Policy, Strategy & Research	C
Chief, Intake Evaluations & Problem Resolution	C
Chief Privacy Officer	
Chief Privacy Officer	N
Deputy Chief Privacy Officer, Privacy	C
Deputy Chief FOIA Officer, Freedom of Information Act	C

U.S. Coast Guard	
Civil Rights and Civil Liberties	
Civil Rights and Civil Liberties Officer	P
Deputy Officer, Equal Employment Opportunity Programs	C
Deputy Officer, Programs and Compliance	C
Executive Officer	C
Office of Counternarcotics Enforcement	
Director	S
Chief of Staff*	C
Principal Asst Director	C

S = Presidential Appointee with Senate Confirmation
P = Presidential Appointee
N = Non-Career SES or Schedule C
C = Career
L = Limited term appointee
T = Scientific Professional
* = First Assistant, pursuant to the Federal Vacancy Reform Act

REFERENCES

[1] HSAC, Council for Excellence in Government, National Academy of Public Administration, and think tanks are examples of partners to assist in message formation and delivery.

[2] "Succession Order and Delegation of Authorities for Named Positions" signed October 25, 2007 by Secretary Chertoff lists leadership positions for each component and the order of succession to each position.

[3] The Succession Planning for Critical Positions initiative asks each component leader to identify a career person who would ensure continuity in the event of the departure of an appointed executive.

[4] As of December 27, 2007, the Department's Headquarters has 177 Senior Executive positions within its ranks. The breakdown of the 177 is as follows: 12 Presidential Appointments (that are mandated by Congress and require Senate approval), 33 political appointees, 107 career employees, and 25 positions currently vacant (of the vacant positions at least 3 would be political appointees). Note: The Department is in the process of recruiting/hiring for the vacant positions. **Note:** For the purposes of this footnote, headquarters is defined as all DHS entities except the seven operational components. The seven operational components are: The Transportation Security Administration, U.S. Customs & Border Protection,

U.S. Citizenship & Immigration Services, U.S. Customs Enforcement, U.S. Secret Service, Federal Emergency Management Agency, and the U.S. Coast Guard.

In: Maintaining Homeland Security
Editors: A. P. Proctor et al., pp. 29-54
ISBN: 978-1-60692-990-2

Chapter 2

TOP TEN CHALLENGES FACING THE NEXT SECRETARY OF HOMELAND SECURITY*

Homeland Security Advisory Council

September 11, 2008
Secretary Michael Chertoff
U.S. Department of Homeland Security Washington, D.C. 20528

Dear Mr. Secretary,

On this September 11 anniversary, as we reflect on that dark day seven years ago and remember those who were lost, we present to you a chapter detailing the ten top strategic challenges that will face your successor. It is the hope of the Homeland Security Advisory Council that this chapter will give the next Secretary insight into the key challenges facing this Department at this time.

During the course of our deliberations, it became evident that this list of challenges should not be presented in any particular priority order. Instead, these key issues are interdependent and equal in importance. Each of these challenges must be addressed and overcome to achieve a more secure, prepared, and resilient America. The challenge for the new leaders will be to make hard political choices between competing and equal priorities. This

---

* This is an edited, excerpted and augmented edition of a Homeland Security Advisory Council publication.

chapter provides a template that may help the new leadership successfully navigate some of those hard decisions that lie ahead.

We appreciate the opportunity to advise you on our most important homeland security matters during this crucial time. On behalf of the members of the Homeland Security Advisory Council, it has been an honor to serve you, the members of this Department, and the American people in this advisory role.

Sincerely,

Judge William H. Webster
Chair, Homeland Security Advisory Council

Secretary
U.S. Department of Homeland Security
Washington, DC 20528

Homeland Security

JUN 0 5 2008

Memorandum

MEMORANDUM FOR: Judge William Webster
Chairman
Homeland Security Advisory Council

FROM: Secretary Michael Chertoff

SUBJECT: Homeland Security Advisory Council Tasking

The Homeland Security Advisory Council has contributed to the homeland security of the United States and the betterment of the Department of Homeland Security through 22 completed reports with more than 300 recommendations since its inception. In an effort to support the first transition of administration for the Department, I request that the members of the Council provide for me a list of the ten most pressing strategic-level challenges that will confront the next Secretary of Homeland Security.

This list of challenges – whether outstanding priority recommendations that have yet to be implemented, or other issues of vital importance to the Nation's security – should be at a very broad, strategic level, providing insight into the key challenges that remain for the Department, highlight potential courses of action available, and serve as an appropriate initial briefing for the new Secretary as he or she continues our efforts to secure the Homeland.

Please deliver this report to me by September 15, 2008.

EXECUTIVE SUMMARY

In an effort to assist with the first Presidential administration transition of the Department of Homeland Security, the Homeland Security Advisory Council has identified ten key challenges that will confront the next Secretary of Homeland Security.

It is important to note that these key challenges are in no particular order of importance. They are interrelated and interdependent, equally necessary to continue to build a strong Department and secure Homeland. With that said, a core test for political leadership will be to make the hard choices on priorities and trade-offs between equally important programs and policies. Successfully taking on these challenges will help the new Secretary gain credibility within the Department as well as among its many homeland security partners throughout the Nation. Ultimately, homeland security is about synchronizing efforts with multiple partners across the landscape of America. The ability to successfully establish and maintain meaningful partnerships at all levels of government and society for the purpose of securing the homeland may be the greatest, ongoing challenge facing the next Secretary, as well as his or her successors.

- Key challenge 1: Homeland security is more than just a single cabinet Department.
- Key challenge 2: Quickly get an inventory of the Department's commitments and deadlines and work with Congress to achieve a rational system of oversight.
- Key challenge 3: Continue to improve intelligence and information sharing.
- Key challenge 4: Build a cadre of homeland security leadership through a unified national system of training and education.
- Key challenge 5: Build the strong research and development, procurement and acquisition process necessary to support the Department's various missions.
- Key challenge 6: The work of strengthening our Nation's disaster response capabilities is not complete.
- Key challenge 7: Lead the building of a resilient America.
- Key challenge 8: Find the right balance between secure borders and open doors to travelers, students, and commerce.
- Key challenge 9: Improve risk management and risk communications for homeland security.
- Key challenge 10: Sustainability of our Nation's homeland security efforts.

Top Ten Challenges

The Department of Homeland Security leads America's efforts in the prevention of, protection from, response to, and recovery from attacks, major disasters or other emergencies within the Homeland. In the past year the Department has assisted state, local, and tribal governments with combating wildfires, recovering from multiple hurricanes and floods, and protecting the homeland from terrorist threats. On a daily basis, the Department's various components and organizations pursue multiple missions working together with the common goal of securing the Homeland.

The Department reached its five-year anniversary this year. It is an important milestone in an unprecedented effort to combine over 30 federal agencies and programs into one organization. During this time, the Department has matured through both failure and success, and the men and women involved in this effort are to be commended for their tremendous dedication and resolve in serving and securing our Nation.

The Homeland Security Advisory Council appreciates the opportunity to take stock of the continuing challenges in homeland security in order to distinguish the ten most pressing issues facing our nation and the Department. The Council offers this document to the next Secretary of Homeland Security with the hope that it provides guidance to ensure a smooth transition to a new Presidential administration by helping to add understanding to this complex and newly emerging profession.

The top ten challenges that will face this leadership fall into four areas: first, the broad homeland security environment; second, the key issues and relationships outside of the Department; third, the key issues and challenges within the Department; and fourth, several broad strategic challenges.

It is important to note that these key challenges are in no particular order of importance. They are interrelated and interdependent, equally necessary to continue to build a strong Department and secure the Homeland. Highlighted here are unranked and concurrent issues that must be viewed as such. With that said, a core test for political leadership will be to make the hard choices on priorities and trade-offs between equally important programs and policies.

Successfully taking on these challenges will help the new Secretary gain credibility within the Department as well as among its many homeland security partners throughout the Nation. Ultimately, homeland security is about synchronizing efforts with multiple partners across the landscape of America. The ability to successfully establish and maintain meaningful partnerships at all levels of government and society for the purpose of securing the homeland may be the greatest, ongoing challenge facing the next Secretary, as well as his or her successors.

I. The Homeland Security Environment

Homeland security provides some of the most complex policy and operational challenges faced by any executive agency. Our Nation's homeland security efforts go to the core of long-standing, purposely designed tensions within our Constitutional system of governance; security versus liberty, federalism in intergovernmental relations, and numerous other complex issues. The next Secretary will face two key challenges in understanding the homeland security environment: homeland security involves far more than just DHS, and DHS has a pre-existing set of complex issues, deadlines, and decisions that must be addressed.

Key Challenge 1: Homeland Security Is More than Just A Single Cabinet Department

The Department of Homeland Security is unique in government in its complex need for both horizontal and vertical integration with other organizations and groups. Securing the homeland requires sharing responsibility horizontally with other federal departments and agencies. Relationships with the Intelligence Community, the Department of Justice, the Department of Defense, and other partners must continue to grow and become institutionalized. DHS must continue to build its ability to serve as a leader in interagency homeland security processes and programs. Homeland security also requires a robust vertical integration of the federal, state, local, and tribal governments, the private and non-profit sectors, and the American citizen to build a secure, safe, and resilient Nation.

While horizontal integration requires traditional leadership at the federal level, vertical integration requires new and collaborative skills that will challenge DHS leadership. Gaining credibility with partners outside of the Department will be the foundation of success for the next Secretary. Key to gaining this credibility is ensuring that a bottom-up approach is applied to virtually all homeland security endeavors. The wide variety of homeland security partners makes this a tremendous challenge.

Building upon and improving existing systems for engagement with the private sector is especially essential. While 85 percent of our nation's critical infrastructure is in the hands of the private sector, and the Department has specifically developed Sector Coordinating Councils to improve private sector coordination in this area, the next leadership team must recognize that private sector infrastructure providers comprise only a small percentage of the entire private sector. The next Secretary must continue to engage the broader private sector to strengthen the working relationships on proposed homeland security strategies, policies, and procedures during their formulation and prior to publication or implementation. This collaboration is particularly important with respect to the private sector's role in our Nation's response and resiliency endeavors, and most importantly with respect to the Department's regulatory rule-making authorities.

The Homeland cannot be secure until every hometown is secure. That means listening to local leaders and including their thoughts in the homeland security plans, programs, and processes. The incoming leadership of DHS will be challenged continually to ensure that homeland security partners are provided collaborative mechanisms to provide essential "outside-the-beltway" perspectives that work towards securing the Homeland.

Key Challenge 2: Quickly Get an Inventory of the Department's Commitments and Deadlines and Work with Congress to Achieve a Rational System of Oversight

DHS is involved in many controversial but important efforts, many with congressionally-mandated deadlines. The ability to competently manage these deadlines will test the credibility of the new Secretary and the new leadership team. Additionally, 86 Congressional committees are keeping a keen eye on the Department's progress in meeting these numerous efforts and deadlines. For example, the 9/11 Act requires that DHS scan 100 percent of U.S.-bound maritime containers by 2012 and screen 100 percent of air cargo by 2010. Both requirements also lay out key milestones that must be met (e.g., the requirement that 50 percent of air cargo be screened by February 2009). The new administration will need to ensure consistency in the Department's approach to meeting both the 100 percent maritime scanning and 100 percent air cargo screening requirements within the 9/11 Act.

The next administration will also need to fulfill requirements for the National Preparedness System. Homeland Security Presidential Directive 8 and the Post Katrina Emergency Management Reform Act (PKEMRA) which call for a comprehensive system to assess, on an on-going basis, the Nation's capabilities and overall preparedness, including operational readiness.

Some additional commitments and deadlines facing the next Secretary and requiring continued action include Real ID, Secure Border Initiative Network (SBINet), US-VISIT, Western Hemisphere Travel Initiative (WHTI), E-Verify, Transportation Worker Identification Credential Program (TWIC), the Target Capabilities List (HSPD-8), and the Electronic System for Travel Authorization (ESTA). This is just a small sample of the numerous programs and activities that the new leadership team will need to tackle quickly.

The new Secretary must also aggressively work with Congressional leadership and The White House to develop a rational system of oversight that provides proper balance between the Executive and Legislative roles in governance and does not choke the ability of the new leadership team to manage the Department. Reporting to 86 different Congressional committees directly and negatively affects the Department's ability to manage and address its many homeland security missions. The next Secretary should continue to pursue the recommendation of the 9/11 Commission for Congress to reduce its homeland security oversight processes. The Congressional system of oversight is a dangerous obstacle to the Department's efforts to achieve its mission.

II. Looking Outside DHS

Building and strengthening partnerships with organizations outside the Department is one of the most fundamental challenges for the new Secretary. Closely linked to this is the need to expand and improve information sharing processes that tie all the players together, and provide a common picture of the challenges and threats to America. Building out a robust system for sharing information and intelligence is a key challenge and underlying network that will tie DHS to all its partners.

Key Challenge 3: Continue to Improve Intelligence and Information Sharing

The Department and its partners have spent the past five years building the basic systems and mechanisms to share and exchange information and intelligence. The next Secretary of Homeland Security must continue to expand this system.

Several opportunities exist to accomplish this goal, including implementing objectively measurable processes for intelligence and information sharing with the private sector, as well as state, local, and tribal governments. The incoming DHS leadership must institute standards to continuously improve the quality, timeliness, and operational utility of intelligence and information-sharing systems with all Homeland Security partners. Several initiatives should be considered and supported as next-steps in building this information and intelligence-sharing network, including:

- **Requirements-based intelligence** – Consistent with the intelligence cycle, DHS should implement a requirements-based intelligence and information sharing process with its state, local, tribal and private sector partners that will allow for each partner to define, identify, and voice its intelligence and information sharing needs.
- **Expand and sustain fusion centers** – Intelligence fusion centers at the state level are becoming an essential place for information sharing. Expand and sustain existing fusion centers by completing the assignment of DHS intelligence officers in every fusion center, continue to build common standards and common platforms for information-sharing in the fusion centers, and seek mechanisms to provide sustained funding to fusion center efforts, especially those efforts that include technology and people.
- **Involve the private sector in fusion centers, emergency operations centers, and joint field offices** – These centers offer the best opportunity for the private sector to integrate with government activities by sharing information on a day-to-day basis and integrating response efforts during times of national security emergencies.
- **Fix the security clearance and classification process** – The federal security clearance process and classification system is broken and is a barrier (and often an excuse) for not sharing pertinent information with homeland security partners. The next Secretary should direct a concerted effort to resolve these clearance and classification issues.

III. Looking Inside DHS

As the new Secretary begins to understand the challenging operating environment of DHS, it is also necessary to look inside the Department and take the next steps to lead DHS towards its 10th anniversary, and beyond. Key challenges inside the Department include building an effective leadership cadre for the future, improving the Department's technology acquisition processes and procurement procedures to prepare for the future threats to the Homeland, and continuing to improve upon the Department's leadership role in preparing and responding to unfolding catastrophic events. While it is inevitable that some continued reorganization of the homeland security enterprise will take place as the new leadership looks inside the Department, the next Secretary should take caution in quickly reorganizing the pieces before gaining a grounded understanding of how the entire organization functions. While some movement will be inevitable, it should be measured against costs and effort that will result.

It is common for new leadership teams to focus on the organizational structures and boxes as the source of their energies. But in reality it is the people making up the organizational charts that make mission successes possible. The next Secretary should focus on the needs of the employees within DHS and the supporting management systems and infrastructures that enable them to accomplish their work. By placing the people inside the organization first and understanding their jobs and roles, any necessary reorganizing of those people into a structure for accomplishing their missions will become self-evident.

Key Challenge 4: Build a Cadre of Homeland Security Leadership through a Unified National System of Training and Education

Homeland security remains in its early stages as both a concept and as a profession. It is a mosaic of disciplines, from law to law enforcement, public policy, administration, military and national security affairs, emergency response, emergency management, risk management, intelligence, science and technology, and communications. Professionals from each of these fields populate the federal, state, local, tribal, private, and non-profit leadership throughout our national homeland security system. It is imperative that the next Secretary work to secure the grounding of this profession by aligning the training and education systems on a national level. This must start inside the Department with its own leadership, and expand outside the Department to include the wide variety of homeland security partners.

- **Leadership training environment** - One initiative the Department should continue to support and expand is the Homeland Security Academy program with the Homeland Security University System to develop its own leadership school where senior and future leaders in the Department can share experiences, standardize their professional development, and work to further integrate the Department's culture, knowledge, and operations. Developing a place where students who are key leaders from multiple DHS organizations can interact, develop relationships, discuss key leadership and cultural issues, and work together on projects that are beneficial to the Department is the key to this effort. Currently, DHS is using the Naval Postgraduate School to support this program. There are many examples in private industry that can be modeled to create a leadership training environment to develop future leaders.
- **Professional development** - DHS must lead an effort to increase the visibility of professional development opportunities both within DHS and with homeland security partners. This will involve expanding the opportunities for training, education, and development assignments beyond traditional horizontal opportunities (to work in different DHS components or other Cabinet departments, for example) to include opportunities to work vertically, with state, local, and tribal governments, or with the private or non-profit sectors. Understanding both the horizontal and vertical relationships – and how these interlocking pieces fit into our Constitutional system of government -- is the key to unlocking the potential of DHS by developing strong future leadership who will lead a truly national homeland security effort.
- **Education** - Over 200 colleges and universities are now providing degrees in homeland security and related fields. For a homeland security degree to mean something, however, people must know what a homeland security degree means. DHS must lead an effort to align curricula, develop education standards, define the loose boundaries of the profession, and support the academic foundation of a homeland security education system. The concept of the Homeland Security University System must be expanded to include a systematic, national approach to homeland security education.
- **National Security vs. Homeland Security education** - It is important that the next Secretary understand and emphasize that "homeland security" is aligned with, overlaps, but is not the same as "national security." Commendable efforts are underway to rebuild the system of National Security Professional Development at the federal level as dictated by the President's May 2007 Executive Order 13434. DHS must ensure that a similar effort, taking into account the unique nature of homeland security, provides the Nation with

leaders for the future with the professional understanding of how homeland security fits into the national security jig-saw puzzle.

- **Training** - DHS has supported many training programs, often through grants, to build levels of operational competency. DHS should ensure that these training programs are reviewed for relevance and align with the rest of the homeland security training and education system.
- **Enhance Red Team Thinking** - DHS should also look for mechanisms to enhance "red team" thinking and capabilities among its leadership by incorporating programs that provide alternate world-views and build the capacity of leaders to think in a more flexible and adaptive manner. Leaders should learn to think like our adversaries, or to simply look at problems through different lenses and test institutional assumptions.

Key Challenge 5: Build the Strong Research and Development, Procurement and Acquisition Process Necessary to Support the Department's Various Missions

DHS must continue to build a credible and reliable system for identifying, developing, and acquiring technologies and services that meet the evolving requirements of the Department and the changing threats to the Homeland. The next Secretary must continue the effort to implement the recommendations from the Homeland Security Advisory Council's *Essential Technology Task Force Report* (2008), including:

- Build a high performance acquisition function implemented by capable staff.
- Adopt a rigorous Department-wide requirements management process.
- Develop a Department-wide acquisition strategy with a clear implementation plan.
- Improve engagement with the private sector in the acquisitions process.
- Manage innovation through a variety of approaches.
- Use the regulatory and standards-setting role of DHS to generate economies of scale across markets.

Key Challenge 6: The Work of Strengthening Our Nation's Disaster Response Capabilities Is Not Complete

The new Secretary will need to understand the evolving roles, responsibilities, and authorities that go along with his or her position as the Nation's leader for domestic incident management. Understanding the relationships between the Secretary of Homeland Security and the roles of the Attorney General, the Secretary of Defense, the Administrator of FEMA, and state and local leaders will test the leadership of the next Secretary. The issue of when and through what authorities the President and the Nation's Governors must share or cede authority during a large catastrophe is still an unsettled Constitutional issue.

Despite good efforts to improve our national response capabilities, including the new National Response Framework, the next Secretary must be aware of many response issues that require additional attention.

- **The gap in medical and EMS surge capacity** – Gaps remain in the Nation's disaster preparedness and response capabilities, most notably in medical surge capacity and emergency medical readiness for mass casualties and specialty life support, trauma, and burn care. This will require close coordination with the next Secretary of Health and Human Services.
- **National Planning System and Target Capabilities** – As national planning efforts are established, the new Secretary will be challenged to ensure involvement of homeland security partners in building a bottom-up approach of organization and response. This planning is essential but must involve homeland security partners to make those plans realistic. Eisenhower's dictum that planning is important, not the plan itself, applies here.
- **Testing of response capacity** – The new Secretary will be involved in many national exercises, an important tool used for determining preparedness and capabilities. However, the exercise system currently in use has many weaknesses, often either throwing predictable scenarios at participants in order to involve everyone or failing to test for system failure to identify areas and issues where the system can be strengthened. DHS should also enhance the use of Red Teams in exercises, not just to probe for weaknesses and overlooked vulnerabilities, but also to test for weaknesses that could threaten response efforts. Additionally, key lessons are more often revisited than learned and the process for sharing lessons from exercises and tying them into program changes is ineffective and slow.
- Implement the National Incident Management System (NIMS) and the National Response Framework (NRF) across the federal agencies – Training partners in

the federal government in the use of the NRF and the NIMS system and implementing its widespread use must become a government-wide priority.

- **Monitor the real-time resource status of national response assets** – DHS has improved its ability to maintain situational awareness during catastrophic events but should work to improve its ability to ensure DHS has real-time visibility on national resources that are available for incident response.
- **Maintain the all-hazards approach** – The all-hazards philosophy should remain an underlying focus of all response and recovery efforts and should also guide the Department's prevention and protection doctrine.
- **Disaster communications** – Communications in the midst of a disaster is a perpetual problem, and DHS must continue to lead initiatives that work to ensure resilient communications systems and interoperability during disasters, as well as establish response plans and mechanisms that can be executed in the midst of, or in spite of, communications failures.
- **Think about recovery before the catastrophe** – DHS should develop a framework for catastrophic disaster recovery in anticipation of future events to take place alongside preparedness, mitigation, and planning as a key inter-governmental function.

IV. Strategic Initiatives

In addition to the previous objectives, there are a number of strategic national challenges that the next Secretary will need to address. These include shifting our preparedness and protection efforts towards a concept of national resiliency, finding the right balance between security and openness at our borders, and building the framework that will support a risk management approach to the homeland.

Key Challenge 7: Lead the Building of a Resilient America

Critical infrastructure is the enabler of our national economic and social activity and therefore, a central focus of our homeland security efforts. While the Department's role in promoting infrastructure protection is important, recent infrastructure failures and their cascading consequences have demonstrated that a focus based solely on protection is not enough. As a nation, we cannot protect everything, against all things, at all times, and at all costs. Fortress America thinking is an unattainable goal and the wrong national strategy. A far-superior objective for our Nation and our critical infrastructures is the concept of resilience. It is a concept that is more aligned with our traditional American values as an independent, rugged, and strong society. The ability to absorb the blows and quickly snap back from the consequences of any event, natural or manmade, will be the measure of long-term security success. The Nation-wide application of a "resilience metric" (i.e., time to reconstitution of every day services and routines of life) builds on traditional, sector-focused protection efforts and provides the means to objectively assess, triage, and significantly mitigate and effectively manage the initial and cascading consequences of infrastructure service disruption, regardless of cause. An all-hazards approach to building resiliency should become an overarching theme throughout our homeland security enterprises.

- **Public-private partnerships** – Resilience is a unifying goal that must be addressed by the entire Federal Government and Nation. Thus, the new Secretary must ensure that DHS policies and programs empower, enable and leverage the experience, vision and innovation that reside in private sector, community, state and regional-based resilience efforts. Ensure continuous improvement in national resilience by actively engaging and acting upon private-sector and academic-sector thought leaders (e.g., Business Executives for National Security, Center for Strategic and International Studies, The Council on Competitiveness, Council on Foreign Relations, The Heritage Foundation, The Infrastructure Security Partnership, and The Reform Institute). As an extension of the above, the government must also support the development of a nationwide system of community-based, cross-sector, resiliencefocused partnerships to empower communities to collaborate in their collective best interests.

Key Challenge 8: Find the Right Balance between Secure Borders and Open Doors to Travelers, Students, and Commerce

This issue involves more than just DHS, including such partners as the Department of State, The White House, and the private sector. The challenge of finding the proper balance between secure borders and open doors for travelers, foreign students, foreign workers, and commerce remains a priority. This issue must be a focus of the highest levels of our government.

- **Need for Cabinet-level leadership** – We acknowledge that there is an inherent and unavoidable tension between security and openness, requiring policy analysis and review by the President and the Cabinet. It is unreasonable to expect mission-oriented agencies within the Department, e.g. Customs and Border Protection or the Transportation Security Administration, charged with a specific security mission, to adjudicate the trade-off between security and openness. But we cannot win the long struggle against extremist ideology by closing our doors to the people of the world that want to visit, learn, and work here.
- **Expand what works** - The Department has made great accomplishments in improving security by identifying those who would do us harm. Expanding programs such as Model Ports is important, and these programs are to be commended. But additional work is needed, especially with inter-agency partners such as the Department of State, to improve the processing of visitor, student, and worker visas. In addition, DHS must continue to work with industry partners to ensure that the flow of commerce and the life of our economy is not choked off in an effort to build walls against both real and perceived threats.
- **Improve metrics** - Key to understanding this problem is the need to further develop and refine better metrics that measure the performance of our security systems around the processing of people and goods. An effort to integrate the metrics used at the Department of State and in private industry with those used at DHS would help build a common picture of the movement of people and goods, and help measure our effectiveness in achieving the right balance.

Key Challenge 9: Improve Risk Management and Risk Communications for Homeland Security

Dealing with risk involves two key concepts: risk management and risk communications. The new Secretary must support and expand efforts to improve the risk-based approach to homeland security, which is in its early stages. The evolution and decisions about risk, as applied to the Homeland, are paramount to building a risk-based foundation for security that lasts into the next decade. Ultimately, a good risk framework is only useful if political leaders (at all levels) are willing to make tough choices on security trade-offs. In addition, the new Secretary has the challenge of communicating to the American people and homeland security partners in an effort to prepare them for potential threats, as well as during unfolding crises.

- **Risk management** – Assessing the changing risks to America is always a priority. DHS is uniquely positioned to lead a national effort at developing a risk management approach to securing the Homeland. Determining the risks to the Homeland and using a risk management approach to allocate resources, make decisions, and communicate threats, readiness, and protective actions has not been perfected. This will require establishing and improving performance metrics for measuring risk and building a framework for risk-informed decision-making. An effective risk-based approach will help the new Secretary share the responsibility for managing those risks with other partners in homeland security, allocate resources more efficiently and effectively, and counter the politicization of homeland security programs (such as grant distribution). It is also important that the Department make an effort to consolidate the different, existing risk management programs across its many components and agencies, to ensure that the DHS risk methodology is consistent within the Department, and consistent when presented to the Department's many partners.
- **Risk communications** – Improving the approach to risk in the Department also requires improving risk communications systems. Risk communications starts with the fundamental need to explain the threats that America faces and involves creating a culture of preparedness through long-term education. An effective risk communications system can mitigate the long-term psychological effects of an incident on the American citizen, which will help safeguard our Democracy against overreaction to possible future attacks. It also involves improving crisis communications systems such as the color-coded homeland security advisory system (which should be revisited and revised to have meaning for the average American), as well as completing the improvements to the old emergency alert and warning systems that DHS controls. New crisis communications networks must engage homeland security partners during the initial phases of an event, provide adequate and immediate warning with clear instructions to those in harm's way, and continually update partners and the public with pertinent information as the situation changes.

Key Challenge 10: Sustainability of Our Nation's Homeland Security Efforts

Two underlying issues challenge the long-term sustainability of our homeland security efforts: finances and focus.

- **Financial sustainability** – In this period of economic struggle, it will be easy for the federal government, state, local, and tribal governments, and partners in the private sector to shirk responsibility for financing long-term homeland security efforts. Selling the need for continued and long-term investments to finance our homeland security enterprise for those over-thehorizon threats will challenge the next Secretary.
- **Focus** – One of the toughest challenges for the next Secretary will be to keep the American public engaged and focused on efforts to prepare for any potential domestic or international threats. Maintaining the political will and public support to move forward with necessary, long-term efforts to protect the Homeland is a challenge for all Americans. The new leaders of DHS must continue to work to inspire a sense of urgency in the public and among policy-makers that our efforts are necessary and valuable to the protection of our Nation.

Acknowledgements

The members of the Homeland Security Advisory Council would like to express our appreciation to the three leaders we have had the privilege of advising: President George W. Bush, Secretary Tom Ridge, and Secretary Michael Chertoff. We believe that their leadership in facing the fundamental challenge of our generation in the wake of the attacks of September 11, 2001 will be seen through the eyes of history as laying the strong foundation for the future security of America.

We have found this Council to provide a unique avenue for exchanging important ideas and initiatives among all the partners who share the burden of securing America. We offer this chapter to the next Secretary with the hope that he or she will have a successful tenure in leading this young organization. This Council stands ready to assist the new Secretary in our continued desire to be of service to the Department and the American people.

The Council would also like to express appreciation to the staff of the Homeland Security Advisory Council whom have worked diligently to facilitate the delivery of our advice and recommendations.

PROCESS

Department of Homeland Security Secretary Michael Chertoff tasked the Homeland Security Advisory Council with compiling a list of top challenges facing the Department at the Advisory Council's meeting in Washington, D.C in June 2008. During that meeting, members heard views on current challenges expressed by members of the Department and its senior leadership, homeland security partners, and homeland security thought leaders.

A working group of Homeland Security Advisory Council members and its five subcommittees held eight separate teleconferences in July, August, and September 2008 to discuss and develop this chapter. The members also reviewed past Advisory Council materials and position papers from leaders in academia, think tanks, the media, Congress, the GAO, and the Department's numerous homeland security partners in government.

APPENDIX A: HOMELAND SECURITY ADVISORY COUNCIL

Homeland Security Advisory Council Members

William H. Webster (Chair)
Partner, Milbank, Tweed, Hadley & McCloy, LLP
James R. Schlesinger (Vice Chair)
Chairman, Board of Trustees, The MITRE Corporation
Richard Andrews, Ph.D.
Senior Director, Homeland Security Projects,
National Center for Crisis and Continuity Coordination (NC4)
Norman R. Augustine
Retired Chairman and Chief Executive Officer, Lockheed Martin Corp.
Kathleen M. Bader
Board of Directors, Textron, Inc.
Elliott Broidy
Chairman and Chief Executive Officer, Broidy Capital Management
Kenneth C. Canterbury, Jr.
President, Fraternal Order of Police
Frank J. Cilluffo
Associate Vice President for Homeland Security,
The George Washington University

Jared L. Cohon
President, Carnegie Mellon University
Ruth A. David, Ph.D.
President and Chief Executive Officer, Analytic Services Inc.
Louis Freeh
Senior Managing Partner, Freeh Group International
Lee H. Hamilton
President and Director,
Woodrow Wilson International Center for Scholars
Glenda Hood
Former Mayor of Orlando and Florida Secretary of State
President and Chief Executive Officer, Hood Partners, LLC
Herb Kelleher
Founder and Chairman Emeritus, Southwest Airlines Co.
Don Knabe
Supervisor, Fourth District, County of Los Angeles
John Magaw
Domestic and International Security Consultant
Patrick McCrory
Mayor, City of Charlotte, North Carolina
Edward Mueller
Chairman and Chief Executive Officer,
Qwest Communications International Incorporated
Erle A. Nye
Chairman Emeritus, TXU Corporation
Sonny Perdue
Governor, State of Georgia
Richard "Rick" D. Stephens
Senior Vice President, Human Resources and Administration,
The Boeing Company
Lydia W. Thomas
President and Chief Executive Officer (Retired), Noblis
John F. Williams, M.D., Ed.D., MPH
Provost and Vice President for Health Affairs,
The George Washington University and Medical Center
Robert L. Woodson, Sr.
President, Center for Neighborhood Enterprise

Hsac Subcommittees

Academe and Policy Research Senior Advisory Committee

Jared L. Cohon (Chair)
President, Carnegie Mellon University
Ruth A. David, Ph.D. (Vice Chair)
President and Chief Executive Officer, Analytic Services, Inc.
David M. Abshire
President, Center for the Study of the Presidency
Sandra Archibald
Dean and Professor of Public Affairs, Evans School of Public Affairs, University of Washington
Daniel I. Goure
Vice President, Lexington Institute
Dr. Victoria F. Haynes
President and Chief Executive Officer, RTI International
Steve Kerr
Senior Advisor, Goldman Sachs
Brian Leary
Vice President, AIG Global Real Estate Investment Corp.
William Parrish
Associate Professor, Homeland Security, School of Government and Public Affairs, Virginia Commonwealth University
Gary M. Shiffman, Ph.D.
Senior Vice President and General Manager, Continuity and Crisis Management, L-3
Roxane Cohen Silver, Ph.D.
Professor, Department of Psychology and Social Behavior, University of California, Irvine
Robert "Rocky" Spane
Vice Admiral USN (Ret.) and Chairman, Halo Maritime Defence
Allan Zenowitz
General Officer and Senior Executive, Retired

Emergency Response Senior Advisory Committee

Richard Andrews, Ph.D. (Chair)
Senior Director, Homeland Security, National Center for Crisis and Continuity Coordination (NC4)
Kenneth C. Canterbury, Jr. (Vice Chair)

President, Fraternal Order of Police
Dan Corsentino
Former Sheriff, Pueblo County, Colorado
Frank P. Cruthers
1st Deputy Commissioner, Fire Department of New York (FDNY)
P. Michael Freeman
Fire Chief, Los Angeles County Fire Department
Ellen M. Gordon
Associate Director, Executive Education,
Naval Postgraduate School, Monterey, California
Arthur Kassel
Chief of Law Enforcement Services,
State of California Department of Mental Health
Phillip E. Keith
Retired Chief, Knoxville, Tennessee Police Department
Paul M. Maniscalco
Senior Research Scientist & Principal Investigator,
The George Washington University
President, International Association of EMS Chiefs
Judith M. Mueller
Director of Public Works, Charlottesville, Virginia
Edward P. Plaugher
Fire Chief (Retired)
Director, National Programs, International Association of Fire Chiefs
Jack E. Reall
President, Fire Fighters Union 67, Columbus, Ohio
Ellis M. Stanley, Sr.
Director of Western Emergency Management Services, Dewberry
Joseph C. White
Senior Vice President of Chapter Organizations,
American National Red Cross
John F. Williams, M.D., Ed.D., MPH
Provost and Vice President of Health Affairs,
The George Washington University and Medical Center

Private Sector Senior Advisory Committee
Richard "Rick" D. Stephens (Chair)
Senior Vice President, Human Resources and Administration,
The Boeing Company

Herb Kelleher (Vice Chair)
Founder and Chairman Emeritus, Southwest Airlines Co.
Nelson Balido
President and Chief Executive Officer, Balido & Associates, Inc.
Stephen Gross
President, BiNational Logistics, LLC
Dr. Doug Huntt
President and Chief Executive Officer,
Burlington United Methodist Family Services, Inc.
Monica Luechtefeld
Executive Vice President, Supply Chain and IT, Office Depot, Inc.
Maurice Sonnenberg
Senior International Advisor, J.P. Morgan
Jean E. Spence
Executive Vice President, Research, Development and Quality,
Kraft Foods, Inc.
George Vradenburg III
Chairman and President, Vradenburg Foundation
Executive Vice President, AOL-Time Warner (Ret.)
Emily Walker
Former 9-11 Commission Member
Bill Whitmore, Jr.
Chief Executive Officer, AlliedBarton Security Services
Houston L. Williams
Principal Owner, Raven Oaks Vineyards and Winery
Jack L. Williams
Former President and Chief Executive Officer, Eos Airlines

State and Local Officials Senior Advisory Committee
Don Knabe (Chair)
Supervisor, Fourth District, County of Los Angeles
Sonny Perdue (Vice Chair)
Governor, State of Georgia
Karen J. Anderson
Past President, National League of Cities
Former Mayor, City of Minnetonka, Minnesota
Michael Balboni
Deputy Secretary for Public Safety, State of New York
Jim Douglas

Governor, State of Vermont
Past President, Council of State Governments
James R. Dunlap
Former State Senator, State of Oklahoma
Judge Robert A. Eckels
Partner, Fulbright & Jaworski, LLP
Peggy Merriss
City Manager, Decatur, Georgia
Karen M. Miller
District I Commissioner, Missouri Boone County Commission
Douglas H. Palmer
Mayor of Trenton, New Jersey
Former President, U.S. Conference of Mayors
David Wallace
Chief Executive Officer, Wallace Bajjali Development Partners, LP
Indiana State Senator Thomas J. Wyss
Representative National Conference of State Legislatures
Joe Zarelli
State Senator, State of Washington

Secure Borders and Open Doors Advisory Committee
Jared L. Cohon (Co-Chair)
President, Carnegie Mellon University
John Chen (Co-Chair)
Chairman, CEO and President, Sybase, Inc.
Frank Cilluffo
Associate Vice President for Homeland Security,
The George Washington University
Sandy Dhuyvetter
Chief Executive Officer and Executive Producer, TravelTalkMEDIA
Thomas J. Donohue
President and CEO, U.S. Chamber of Commerce
Roger Dow
President and Chief Executive Officer, Travel Industry Association
John Engler
President, National Association of Manufacturers
Susan Ginsburg
Director, Mobility and Security Program, Migration Policy Institute
J.W. Marriott

Chairman and Chief Executive Officer, Marriott International, Inc.
James May
President and Chief Executive Officer,
Air Transport Association of America, Inc.
Maria Luisa O'Connell
President, Border Trade Alliance
Eduardo J. Padrón
President, Miami Dade College
Greg Principato
President, Airports Council International-North America
James A. "Jay" Rasulo
Chairman, Walt Disney Parks and Resorts
Keith Reinhard
President, Business for Diplomatic Action
Chairman Emeritus, DDB Worldwide
Paul B. Roth, MD
Executive Vice President, University of New Mexico
Health Sciences Center Dean, School of Medicine
Roxane Cohen Silver, Ph.D.
Professor, Department of Psychology and Social Behavior,
University of California, Irvine
Richard "Rick" D. Stephens
Senior Vice President, Human Resources and Administration,
The Boeing Company
Rose Mary Valencia
Director, Visa and Immigration Services Administration,
The University of Texas M.D. Anderson Cancer Center
Charles Vest
President, National Academy of Engineering

Executive Director
Jeffrey D. Stern

Directors
Charles J. Adams
Mike Miron
Candace Stoltz

Staff

Jennifer R. Myers

Amanda Rittenhouse, DHS Honors Fellow

APPENDIX B: HOMELAND SECURITY ADVISORY COUNCIL REPORTS

2008

- Report of the Essential Technology Task Force »
- Report of the Administration Transition Task Force »
- Report of the Secure Borders Open Doors Advisory Committee »

2007

- Future of Terrorism Task Force Report »
- Homeland Security Culture Task Force Report »

2006

- DHS/DOJ Fusion Center Guidelines »
- Critical Infrastructure Task Force Report »
- Weapons of Mass Effect Task Force on Preventing the Entry of Weapons of Mass Effect Into the United States »

2005

- Homeland Security Requirements Process »
- Homeland Security Information Sharing between Government and the Private Sector Report »
- Homeland Security Intelligence and Information Fusion Report »
- Maritime Strategy Report »
- Homeland Security Presidential Directive 8 Recommendations »

2004

- Intelligence and Information Sharing Initiative Report and Recommendations »

- Task Force on State and Local Homeland Security Funding Report »
- Homeland Security Lexicon Recommendations »
- Homeland Security Award Recommendations »
- Statewide Template Initiative »

In: Maintaining Homeland Security
Editors: A. P. Proctor et al., pp. 55-78
ISBN: 978-1-60692-990-2

Chapter 3

ESSENTIAL TECHNOLOGY TASK FORCE*

Homeland Security Advisory Council

June 25, 2008
Secretary Michael Chertoff

U.S. Department of Homeland Security Washington, D.C. 20528

Dear Mr. Secretary,

I am pleased to present to you the final report of the Homeland Security Advisory Council's Essential Technology Task Force.

Per your direction, the Homeland Security Advisory Council established the Task Force to provide recommendations for acquiring technologies critical to meeting the Department's mission of maintaining a robust, flexible, and resilient homeland security capacity when both threat and technologies are rapidly changing. As directed, the Task Force began assessing the utility of using alternative legal and financing tools to acquire the essential technologies by consulting key subject matter experts from the public and private sectors.

During the course of the Task Force's research and deliberation, it became clear that, as a precondition for employing alternative financing tools, the Department must improve its overall requirements and acquisition

* This is an edited, excerpted and augmented edition of a Homeland Security Advisory Council publication.

processes so that it has the capability of assessing the merits of varying legal and financial mechanisms. The Task Force believes that when a robust, comprehensive acquisition strategy is in place, DHS will have the necessary internal capability to effectively decide among the legal, financial, and operational options associated with alternative acquisition approaches.

The findings that led the Task Force to make its recommendations are laid out in the attached chapter, which the HSAC has reviewed and approved.

- Recommendation 1: Build a high performance acquisitions function implemented by capable staff.
- Recommendation 2: Adopt a rigorous Department-wide requirements management process.
- Recommendation 3: Develop a Department-wide acquisition strategy with a clear implementation plan.
- Recommendation 4: Improve engagement with the private sector in the acquisitions process.
- Recommendation 5: Manage innovation through a variety of approaches.
- Recommendation 6: Use the regulatory and standards setting role of DHS to generate economies of scale across markets.
- Recommendation 7: Continue to advocate for the reduction of Congressional committees overseeing DHS.

The Task Force also received several briefings from the Transportation Security Administration (TSA) on alternative approaches to financing in-line baggage screening systems and innovative ways of improving passenger screening. Although these briefings did not fall within the ultimate scope of the tasking, the Task Force was impressed with TSA's forward thinking initiatives and endorses both the recommendations of the prior Baggage Screening Investment Study, as well as the newly established Checkpoint Evolution, an adaptive approach to airport security.

On behalf of the Homeland Security Advisory Council membership, thank you for entrusting us with tackling an issue fundamentally important to the Department and crucial to securing our Nation.

Sincerely,

Judge William H. Webster
Chair, Homeland Security Advisory Council

BACKGROUND

The post-9/11 homeland security environment requires an innovative, resilient system of operational capacities, broadly implemented across the public and private sectors. These systems should be capable of identifying and deploying the defensive measures needed within the United States, as well as beyond our borders, to secure the Nation, protect the American people from threats, and to quickly restore businesses and communities damaged by manmade or natural disasters. Technology, itself subject to rapid changes, is a critical element of our homeland security systems.

Within this context, Secretary Michael Chertoff tasked the Homeland Security Advisory Council with establishing an Essential Technology Task Force. The Task Force began by considering how to improve the Department's capability to identify, acquire, and deploy essential technologies to support homeland security needs. It evaluated the strategic, organizational, and operational steps necessary to develop this capability, and the feasibility of alternative acquisition processes and programs for acquiring essential technologies. The Task Force commends the Secretary and the Department for asking the hard question: What can DHS do to improve its performance in the acquisition of essential technologies needed to support the homeland security mission? But before DHS can fully address the viability of alternate approaches in variable and heterogeneous acquisition contexts, it is necessary to first mature the Department's internal acquisition environment to include strengthening its overall requirements and acquisition capabilities in the manner recommended in this chapter.

The Task Force recognizes that critical stakeholders in homeland security include state, local, and tribal partners, and the private sector. Since many essential technologies will be used in partnership with these stakeholders, they have a key role in the technology lifecycle, from developing requirements to implementation and sustainability.

Unlike national defense systems, homeland security systems are frequently installed and integrated within domestic business and consumer environments, presenting novel and complex issues of coordination and collaboration unique to domestic security challenges. Involvement of businesses and the public in developing new approaches to homeland security challenges will improve the implementation of technologies (from strategic to the most tactical levels), the adoption, acceptance, and integration of new systems, and the reduction of barriers working against our homeland security efforts. Involving these stakeholders in early development of new products and systems is critical to a smooth and effective deployment of homeland security solutions. In the end, the American people are the ultimate stakeholders.

The Department is making progress in identifying national needs in today's security environment. To meet its mission, DHS is adapting its management practices and absorbing essential technologies into its programs. This progress includes an update of the February 2004 *Homeland Security Strategic Plan* and the Department's Deliberate Planning Process to drive budgeting as well as the hiring, training, and retention of acquisition and management staff.

Yet, much more must be done to correct what observers both inside and outside DHS view as systemic weaknesses in the Department's organizational agility and operational performance when implementing new technologies.

Notwithstanding the shortcomings we have identified in the course of our work, the Task Force offers this chapter with deep respect for the commitment of the many homeland security professionals we were privileged to meet throughout the course of this effort.

PROCESS DESCRIPTION

The Task Force met in person and by teleconference on multiple occasions between September 2007 and May 2008. Task Force members shared their own public and private sector experiences in technology development, transactions, and transformations. Members also sought the views of public and private sector subject matter experts, many of whom are or have been participants in contracts involving technology acquisitions for DHS and other federal agencies and departments. We also heard the views of staff representatives of Congressional committees, the Government Accountability Office (GAO) and of the Office of Management and Budget (OMB). Based on this input, the Task Force developed a set of **findings** that form the basis for the following **recommendations** concerning a department-wide acquisitions program. The results are presented to the Secretary for consideration.

FINDINGS

The Department is suffering from systemic weaknesses in its ability to accept and manage change in technologies essential to its missions. Many of these ongoing weaknesses have been identified in GAO reports, Congressional oversight hearings, and media reports. A contributing factor is that after five years in existence, the Department does not have a homeland security strategic planning process that provides a coherent, objectively measurable, and repeatable method for setting program and budgetary priorities and trade-offs among the long list of possible desired operational end states. The DHS strategic plan must have clearly defined objectives with tactical, measurable steps for achieving those objectives.

Without a comprehensive strategic plan, DHS lacks robust, department-wide methods for establishing requirements and priorities across multiple missions and components. As a result, major stakeholders in homeland security, to include the public, Congress, GAO, OMB, and the contractor community, lack confidence in the Department's ability to effectively set clear priorities and develop multi-year programs and budgets. Until DHS develops an effective requirements identification, validation, and prioritization process, it is unlikely to see the more flexible budgetary authority afforded to other federal departments. The Task Force acknowledges the Department's current efforts to develop such a process, but in order for it to be fully effective, a comprehensive and robust acquisition process must complement it.

There are accepted standards, processes, and principles that DHS can adopt to make technology acquisition more effective now and in the long run. A disciplined distribution of financial, program execution and mission risk between DHS and its vendors will provide the Department with cost or operational advantages in particular acquisitions.

Currently, DHS is not effectively using acquisition options, including leases and contract services, permitted by existing legal authorities and the Federal Acquisition Regulations. Only with robust requirements and acquisition capabilities and processes in place will the Department be better able to analyze different procurement tactics and trade-offs in large-scale department-wide acquisitions, including capital investment programs, strategic sourcing, services contracts, and grants.

Having the right people, processes, tools, and training in place within DHS is key to having the capacity to successfully execute department-wide technology projects. All four of these elements are indispensable components in building this capacity within DHS. The Task Force understands that developing this capacity will require a multi-year effort by DHS leadership, fully supported by the OMB and Congress, in collaboration with appropriate departments and agencies at all levels.

People

DHS simply lacks the number of people needed department- wide to manage the acquisition process of procuring, deploying, and executing programs and technologies. Although the establishment of the Acquisition Program Management Division (APMD) in the Chief Procurement Office is a solid start, APMD is currently understaffed and will not be fully staffed until FY 2010. This includes determining measurable requirements, developing, and executing an acquisition strategy, developing the appropriate contract vehicles and incentives to support the acquisition strategy, overseeing implementation, monitoring contractor performance, and making adjustments to deal with changing mission requirements and threats. While some of the component agencies have made great strides in this area (such as the Coast Guard's efforts to systemically correct inadequacies in the early life of its Deepwater program), the Department needs to hire and develop the right number of people, across all components and headquarters, who have the ability and the experience to support all aspects of complex acquisitions. While it must be noted that this problem is not unique to DHS but plagues the entire federal government, the problem is particularly acute to DHS because of its relative early stage of organizational maturity and the accelerated demand for homeland security solutions.

Processes

DHS is currently developing a Deliberate Planning Process. The Department must ensure that this program is linked to its new Quadrennial Homeland Security Review (QHSR) and its new Office of Net Assessment (ONA), which are both under development. Until these pieces are in place and linked, DHS acquisition will remain hampered, lacking an overarching process that ties present budgetary expenditures to future requirements. Ideally, the ONA will identify and review long-range future threats and vulnerabilities and compare them to present capacity, thereby developing an ongoing, robust gap analysis. That gap information would then be fed into a QHSR process, aligning the Department's mid-range strategic plan. This process would lay the groundwork for creating and requesting budgets that support bridging both present and anticipated operational gaps with focused initiatives that are both material (e.g. programs or projects) and non-material (e.g. changes in doctrine or training).

Tools

The Department needs to develop and field the right tools to perform and manage a sophisticated life-cycle acquisition process. These tools include procurement, systems engineering, logistics, test and evaluation, and financial analysis. With the right tools in place, DHS can better pursue alternative acquisition approaches such as multi-year budgeting, franchise funds, and "fly-on" competitive funding models for research and development in which contractors are provided research and development funds through a competitive arrangement. An example of the "fly-on" model is the Air Force's development of fighter planes, where money is provided to several contractors to develop prototypes and then a winner is selected after a competition. Winning contractors get additional funding to implement successful, working solutions. This competitive model might work for some DHS technology needs if selectively applied and justified.

Despite substantial resources, including regulatory and standards-setting capabilities, the Department has been unable to exercise its leverage to generate scalable commercial markets for security products and services, which could reduce costs and increase the utility and reliability of the products purchased by DHS and its homeland security partners. For example, airports and seaport security provide ample opportunity to develop large scale contract vehicles that reduce costs. Currently, each airport and seaport is an individual buyer of goods and services. By aggregating requirements and setting standards, DHS enlarges the market for the goods and services it procures, creating economies of scale. The challenge will be to develop these standards and contract vehicles in a manner that still allows for the flexibility to adapt standard goods and services in heterogeneous contexts and to make adjustments midstream as both threats and the technologies change. This approach is particularly powerful in department-wide services contracting and in strategic sourcing of major technology acquisitions.

Training

The lack of trained and experienced DHS acquisition personnel creates an inability to adequately supervise private sector system integrators or to communicate with the program's intended operators throughout the program's conception, development, and support phases. This leads to inconsistent and sometimes ineffective program execution, as well as program delays, cost overruns, and dissatisfied operators. To address this weakness, DHS should better utilize existing government training programs while developing its own training system for acquisition professionals. This is necessary to ensure a steady stream of competent, motivated individuals who are adaptive within a changing technology environment and can effectively do this important work.

The Task Force commends the current centralized acquisition training initiative underway by the Chief Procurement Office's Acquisition Workforce branch. The initiative capitalizes on existing acquisition training sources such as the Federal Acquisition Institute and the Defense Acquisition University (DAU). Additionally, the DHS program offers a complement of acquisition training provided by commercial vendors and through a unique partnership with DAU. Paramount to the success of this initiative, however, is ensuring that needed training resources are protected from reallocation to non-training programs.

While not equipping itself with the right people, processes, tools, and training has created acquisition challenges for the Department, the Task Force recognizes that there are also external impediments to the development of robust DHS requirements and acquisition processes; most significantly the complicated legislative oversight environment in which the Department operates. Congress's complex and conflicted authorization and appropriations process, with 86 Congressional committees asserting varying degrees of oversight over DHS, has impeded the Department's ability to develop the coherent department-wide processes identified above.

RECOMMENDATIONS

1. Build a High Performance Acquisitions Function Implemented by Capable Staff

DHS must continue to build, and Congress must fund, a capable, integrated acquisitions process at DHS headquarters and across components, executed by the appropriate number of personnel with supporting career fields expertise. These fields should include program management, systems engineering, logistics, contracting/procurement, cost estimating, business and financial management, testing, and evaluation. Procurement functions should seamlessly integrate into this acquisition model to ensure that procurement actions have the benefit of robust examination. This can only be accomplished by all the above disciplines working together.

- Currently, requirements for various department-wide programs, including their technology elements, are not well defined. As a result, contractors often are uncertain of what is required. This heightens the risk to the contractors of program changes after procurements are initiated, with corresponding cost overruns and schedule delays. DHS should consider utilizing post-procurement "award teaming" sessions (similar to the DAU "Program Startup Workshop") with contractors who win major DHS solicitations to mutually confirm implementation strategy. DHS should also assess performance of acquired technologies throughout the acquisition process by use of robust testing and evaluation processes.
- DHS should perform a systems analysis of its acquisition processes, from concept generation through fielding and sustained operations, to include the requirements generation and the legislative budget evaluation processes. Based on the results, DHS should strive to develop a family of standard acquisition mechanism processes to include capital investment programs, departmentlevel service contracts, strategic sourcing, grants, and interagency agreements for general use. The Department should also develop computerized modeling tools to test and evaluate multiple combinations of acquisition mechanisms, procurement strategies, and budgeting profiles for contemplated major acquisitions. This will assist DHS in determining an optimal acquisition approach for major efforts.
- A professional, well-staffed, and well-managed acquisition organization within DHS and its components is necessary to professionally manage the wide range of acquisition mechanisms that DHS uses. As discussed earlier, DHS has only recently started to increase the size and sophistication of its internal acquisition capability. This build-up should be a priority, accompanied by the development and use of assessment and accountability tools, such as the periodic reporting of acquisition mechanism status. These tools will enhance the internal acquisition community's ability to effectively execute its responsibilities.

In addition to the Centralized Training Program of the Chief Procurement Office's Acquisition Workforce branch mentioned above, DHS should also consider the following to build and further educate a qualified staff:

- Developing a unified Acquisitions Office reporting to the Under Secretary of Management to provide direct support to the Under Secretary's role as the Department's Chief Acquisition Officer.
- Supporting the Office of Management and Budget and the Office of Personnel Management in their efforts to quantify the adequate number of acquisition staff

needed, and to facilitate improvements in the vetting of DHS requirements and acquisition budgets.

- Utilizing the services of the Federal Acquisition Institute and the Defense Acquisition University for training and studies to create acquisition aids such as computer models.
- Utilizing experienced acquisition professionals from other agencies as mentors to share knowledge and ideas.
- Consider establishing field offices in less-costly communities outside of Washington, D.C. staffed with new hire procurement workers that are trained and led by an experienced cadre.

2. Adopt A Rigorous Department-Wide Requirements Management Process

DHS must integrate requirements and capabilities considerations at higher levels of management. The Department can do this by adopting a collaborative joint requirementscapabilities function managed by DHS headquarters to oversee department requirements and derivative acquisition strategies across DHS missions and components. To properly establish capability requirements, there must be collaboration between the operational user and the management authority framed within a concept of operations.

- DHS should have a centralized authority to manage an overarching capability strategy that ensures that there are no gaps or overlaps in the capabilities portfolios that the Department establishes to support mission objectives. This body would determine the content of the capability portfolios supporting the Department's five mission areas and determine the appropriate system trade-offs across its seven components. This would allow optimal use of finite department resources and reduce duplication within capability areas shared across the Department's components. For example, Customs and Border Protection's Automated Customs Environment (ACE) is a technology platform with potential application across other components and missions. Another example is the Coast Guard's acquisition of the airborne radar and C3 systems to detect aircraft and vessels for its Deepwater Program. These systems may have direct application to Immigration and Customs Enforcement's air and marine interdiction operations. The Task Force acknowledges and endorses the Department's current efforts to establish such a strategy by incorporating the efforts of various offices into one streamlined approach.

- This centralized approach will also allow DHS to define explicit criteria for inserting capabilities and technologies across components and missions. This approach has the potential to strongly incentivize potential bidders, who would be presented with an increased market for a given product due to its multiple applications across components while meeting departmental-wide goals. These goals could include bidders who design less expensive, flexible systems that remain cutting edge yet are capable of software upgrades, with components that employ modular, scalable designs. DHS should also maintain flexible strategies that enable changes in program, products, or vendors in order to effectively respond to changes in the threat environment, changes in the competitive landscape, or advances in technology. This approach would also permit DHS management to set acquisition strategies that employ the full range of acquisition mechanisms, including purchase, leases, grants, capital investment programs, and enterprise service contracting as appropriate, in order to achieve best value and increase the life and flexibility of acquired Departmental capabilities.

3. Develop a Department-Wide Acquisition Strategy with a Clear Implementation Plan

DHS should produce a departmental acquisition strategy with a clear implementation plan that has a definitive timeline for execution. Together this strategy and plan should allow for rigorous management, priority-setting, and budgetary decision-making across missions and components. This strategy should be dynamic, include a variety of program life spans, and provide for changes in threat and hazards environments. The Task Force commends the Department for moving forward with the Quadrennial Homeland Security Review (QHSR) and the Office of Net Assessment (ONA), which should be linked to a department-wide capability/acquisition strategy.

- The current DHS Strategic Plan (February 2004) organizes department goals into five mission areas1 distributed across seven components2. These goals are further broken down into statements of desired end states. This current strategy does not provide for enhanced decisionmaking with regards to a framework of priorities, solutions, or technologies that cross component and mission areas, and vary through time and threat environment. DHS is in the process of updating the current strategic plan and should incorporate such a framework for decision-making. Each DHS component has unique mission requirements. As a result, goals and related programs organized by missions or components are not effectively integrating many technologies that provide capability across multiple areas. It also may not effectively leverage the technology and capability assessments or procured products of other departments for use in complementary missions. Simply put, an acquisitions process based on the current strategic plan will not effectively capture cross-mission or cross-component capabilities.

- A comprehensive, integrated strategy will enable DHS management to more effectively prioritize programs driven by event, time, threat, or funding and gives decision-makers the ability to make understandable trade-off decisions between systems, programs, or technologies. This strategy will also allow for requirement decisions based on cycle time, lease, purchase, or service provisions. An example of an integrated, comprehensive strategy is IBM's *Global Movement Management; Strengthening Commerce, Security and Resiliency in Today's Networked World* authored by IBM Global Business Services3. Another is the Department of Defense's Office of the Under Secretary of Defense for Acquisition, Technology and Logistics' *Strategic Goals Implementation Plan (2008)*4. The DHS QHSR, currently under development, should provide the context in which to develop such a comprehensive acquisition strategy. The Department should also work to implement the Homeland Security Management System set forth in the President's October *2007 National Strategy for Homeland Security*. The second step in the Homeland Security Management System is a strategy that takes into account underlying assumptions, context, tradeoffs, and necessary resources to achieve policy goals.
- Once DHS produces such a strategy and implementation plan, the Task Force believes that government stakeholders and oversight authorities will gain confidence in the Department's ability to build programs and propose budgets based on mission, operations, time and risk-based priorities. Once confidence is gained, oversight authorities may grant the Department a more flexible budgetary authority, especially for multi-year funding, which is a necessity for more competitive pricing on large, longterm procurement contracts as well as the leeway to deal with evolving homeland security threats.

4. Improve Engagement with the Private Sector in the Acquisitions Process

The Department should engage the private sector beginning in the research and development arena through all phases of the acquisition process; from concept exploration through the lifecycle of field implementation. The Department should pursue systematic engagements with private sector experts and key allies in all these areas to identify alternative approaches to solutions and find a "best fit" to DHS mission requirements.

- The key in requesting private sector input in the DHS requirements process is to establish a collaborative partnership that facilitates a team building mentality. DHS leadership should not allow an "us versus them" mentality to interfere with the process. Potential innovative techniques might include rotational assignments where DHS representatives "embed' in contractor facilities, and other deliberately interactive techniques. The goal should be a "win, win" outcome for both parties while maintaining DHS control and oversight over the process.
- DHS should make greater use of the Request for Information (RFI) process in the Federal Register. The RFI should include lifecycle cost estimates, financing terms, and the Department's goal of maintaining program flexibility in an ever-changing threat environment. This process will allow DHS to solicit input from the private sector in a government controlled environment on issues such as appropriate technology for DHS mission needs as well as the optimal acquisition approach for both DHS and the contractor for specific types of large capability acquisitions.
- When working with the private sector, DHS has many acquisition options. A comprehensive acquisition strategy will assist DHS in making long range plans to ensure it has the budgetary flexibility to finance assets appropriately. DHS must invest in the people and processes necessary to drive standardization of procurement processes which provide win-win solutions for DHS and prospective vendors. The following are just a few financial arrangements the Department can implement to meet mission requirements:

  - Leasing with Defined Termination - DHS must be able to define what will happen to the vendor's equipment and/or services once the contract has ended. This will allow vendors to provide the most advantageous price when knowing there is no risk of equipment being taken out of service at the end of the option period or contract. Such a defined "exit strategy" will create confidence from industry and increase the probability of investment in DHS projects.

- Bundled Services with Defined Termination – DHS could also achieve savings by having vendors provide equipment as part of a bundled service. DHS would receive the benefit of "bulk buying" while the vendor would maintain ownership of the equipment.
- Government Provided Equipment - The financial cost to the government of buying equipment should be lower than for a private sector owner. Unlike a private vendor, as a governmental entity, DHS pays no interest on appropriated funds used to purchase capital.

- The Task Force acknowledges that DHS has made attempts to work with the private sector through alternative approaches in its acquisitions process and commends such innovative thinking. For example, in September 2006, the Transportation Security Administration's Aviation Security Advisory Committee unanimously endorsed the recommendation of the Baggage Screening Investment Study (BSIS) established to find cost-effective, timely ways of installing checked baggage screening systems consistent with Congressional mandates and capable of implementation in the heterogeneous public/private US airport environment. BSIS recommended creating a voluntary $3 billion tax credit bond program in which airports would issue debt to pay for infrastructure and baggage handling systems. The Federal government's contribution would be tax credits instead of direct expenditures. The Task Force endorses the BSIS recommendation and others that allow for creative and alternative financing solutions. DHS Transportation Security Administration had planned to put this recommendation into action this year. Unfortunately, the plan is currently at a standstill, awaiting final review and approval from Congress.
- It is also important to note that mission requirements can often be met by the innovative use of non-material solutions, such as shifts in training or doctrine, or by modifying existing systems. These less-costly options should be fully investigated, in concert with the user community, before the Department requests a new system. The end goal is to deliver capability to the user. DHS acquires capacity as part of managing the lifecycle of capabilities. Technology is an enabler, not the end goal. DHS should manage the technology value chain to ensure that acquired technology matches the needs of the user.

5. Manage Innovation through a Variety of Approaches

The Department should utilize varying approaches to maintain visibility and access to essential technologies to achieve homeland security missions. Once DHS is able to clearly identify program requirements to meet mission goals, it should work with DHS Science and Technology (S&T), the private sector, university laboratories, and foreign partners to develop and acquire the required technology through the comprehensive acquisition strategy recommended above.

- The Department should be more agile in its approach to innovation. Significant effort should be made to meet stated mission requirements by identifying and adapting existing government owned technologies or commercial off-the-shelf (COTS) systems. For example, use COTS software, leveraging commercial updates and support. Adapt processes to the COTS software instead of developing software for existing processes. Through private sector collaboration, DHS should identify commercial systems overseeing the movement of goods, people, information, and related security features. Many of these existing systems inherently address commercial concerns that parallel homeland security vulnerabilities, and are robust, resilient, and efficient. If these capabilities cannot be found, only then should DHS undertake new research and development to meet the requirements.
- The Department should consider the example of S&T's Commercialization Process for department-wide application. This process observes that "the private sector is willing and able to use its own money, resources, expertise and experience to develop and produce fully developed products and services for DHS." This plan calls for DHS to provide the private sector with (1) detailed operational requirements and (2) a conservative estimate of potential available markets. For example, under Homeland Security Presidential Directive 8, "Federal departments and agencies that support the purchase of first responder equipment will coordinate their programs with the Department of Homeland Security and conform to the same standards." By providing detailed operational requirements to the private sector for a given piece of equipment, DHS is opening up a market of millions of first responders to the private sector.
- DHS should test and evaluate multiple low-rate production alternatives in actual operating environments. Such an evaluation enables the Department to gain confidence in the technical functionality of a given product, a more informed competitive award process, as well as the effectiveness of the intended operational architecture for its intended use.

- The Department should create an environment that allows DHS users to familiarize themselves with emerging technologies to better meet mission requirements:

  - Build and maintain a hands-on technology site where key officials, program managers, and users can examine emerging technologies.
  - Evaluate emerging technologies against anticipated requirements in an operational environment.
  - Map next-generation enhancements to existing technologies.
  - Deepen partnerships with technology-oriented colleges and universities through the "center of excellence" approach.
  - Identify visionaries, inventors, and inventions with whom or with which DHS might develop sustained engagement.

- DHS S&T should continue its participation in or more fully engage existing intergovernmental technology organizations including:

  - The Intergovernmental Technology Board.
  - The Technical Support Working Group program under the Combating Terrorism Technology Support Office – Department of Defense and Department of State.
  - NATO's Program of Work on Defense Against Terrorism (PoWDAT).
  - In-Q-Tel.

- The Task Force reaffirms the recommendations of the DHS S&T study on the use of venture capital business practices. The study *Venture Capital Concept Analysis*5 was completed by the Homeland Security Institute, with the results documented in its Final Report, December 2005.

6. Use the Regulatory and Standards Setting Role of DHS to Generate Economies of Scale Across Markets

DHS is in a position to influence standards for security products in order to generate larger markets, both nationally and internationally, and in turn decrease overall costs for homeland security products, services, and programs. Formal adoption of current standards or creating new ones, if necessary, will act as a multiplier for goods and services in the homeland security marketplace. Where DHS is a receiver of information security technology, DHS should participate in private sector standard-setting bodies to seek to assure that private sector standards allow for compatible variations needed to meet potentially more robust homeland security implementations.

- A good example of allowing standards setting to positively affect the market is the DHS Science and Technology's comprehensive program. It manages standards across components for required technology, equipment, and systems. The program, which includes international standards, has positively impacted the competitive market by increasing market size, quality, and facilitating international competition, leading to lower prices.
- A department-level standards program should include a deliberate feedback of information at the strategic policy level to address differing requirements in similar markets in order to facilitate standardizing equipment and systems, increasing market size, and ultimately lowering prices.
- The DHS Safety Act offers legal liability protections and is essential for businesses that provide homeland security services and products. Liability is often a greater concern for a contractor than profit. Reduction in contractor liability will reduce a barrier to competition for businesses, especially small businesses, which have solutions for homeland security. An international equivalent to the Safety Act is necessary to energize homeland security markets in partner nations.

7. Continue to Advocate for the Reduction of Congressional Committees Overseeing DHS

One of the biggest hurdles to the Department's ability to mature and effectively address its many homeland security missions is Congress's inefficient and conflicting oversight process. Eighty-six Congressional committees and subcommittees with homeland security oversight authorities is inherently inefficient (for both Congress and DHS) and counterproductive. The cumbersome and unwieldy oversight of DHS by the legislative branch cripples the Department's effectiveness in a number of areas, including requirements development, acquisition, and budgeting. This Task Force reaffirms the report of the 9/11 Commission recommending that Congress should reduce its homeland security oversight process to an authorizing committee and a subcommittee on DHS appropriations in both the House of Representatives and the Senate. This streamlined process would allow for clear and straightforward channels of authority to which DHS leadership can answer, allowing them to focus more on the operations of securing the homeland and less on reporting to a myriad of Congressional committees and subcommittees.

- The narrow focus of these 86 committees, many with conflicting legislative priorities, notably hinders the efforts of DHS to develop a department-wide strategy for acquisition and information technology.
- Thousands of DHS work hours are redirected from addressing the operations of the Department towards responding to Congressional requests for hearings and briefings. From January 2007 to August 2007, the Department participated in 159 hearings and nearly 1,800 briefings. On several occasions, DHS witnesses were called to testify before multiple committees on similar topics.
- The Task Force believes the Executive Branch can and should advance the case for why the current Congressional committee structures are having an adverse impact on homeland security. The solution must come from a partnership with the Executive Branch, the Department, the Office of Management and Budget, and the legislative branch. One possible solution would be a DHS "caucus" with staff to consolidate briefings and collectively engage Congressional stakeholders in DHS strategy and needs.

ACKNOWLEDGMENTS

The Task Force recognizes that this chapter would not have been possible without the cooperation and support of numerous individuals and a quality HSAC support staff. The input of subject matter experts from both the public and private sectors was invaluable, as were the briefings from and discussions with Congressional staff and government professionals from the Government Accountability Office, the Office of Management and Budget, and DHS components and directorates. The Task Force also commends the outstanding efforts of the Homeland Security Advisory Council staff for the planning and support of Task Force meetings as well as the production of this chapter.

Appendix A: Essential Technology Task Force Members and Staff

Task Force Members

George A. Vradenburg III
President, Vradenburg Foundation
Private Sector Senior Advisory Committee (Task Force Chair)
Joseph White
CEO, American Red Cross, St. Louis, MO
Emergency Response Senior Advisory Committee (Task Force Vice Chair)
Dr. Richard Andrews
Senior Director, National Center for Crisis and Continuity Coordination
Homeland Security Advisory Council
Nelson Balido
President and CEO, Balido & Associates
Private Sector Senior Advisory Committee
Elliott Broidy
Commissioner, Los Angeles City Fire and Police Pension Fund
Homeland Security Advisory Council
Dan Corsentino
Former Sheriff, Pueblo County, Colorado
Emergency Response Senior Advisory Committee
Dr. Ruth David
President and CEO, Analytic Services, Inc.
Academe & Policy Research Advisory Committee
Homeland Security Advisory Council
Stephen M. Gross
President, BiNational Logistics, LLC
Private Sector Senior Advisory Committee
Dr. Victoria F. Haynes
President, Research Triangle Institute
Academe & Policy Research Senior Advisory Committee
Phillip E. Keith
Former Chief, Knoxville Tennessee Police Department
Emergency Response Senior Advisory Committee

Stephen Payne
President, Worldwide Strategic Partners and Worldwide Strategic Energy
Secure Borders & Open Doors Advisory Committee
Rick Stephens
Senior Vice President, Human Resources and Administration, The Boeing Company
Homeland Security Advisory Council
Dr. Lydia C. Thomas
President and CEO, Noblis
Homeland Security Advisory Council
David Wallace
Mayor, Sugarland, Texas
State & Local Senior Advisory Committee
Houston L. Williams
Principal Owner, Raven Oaks Vineyard and Winery, Los Gatos, CA
Allan Zenowitz, BG (Ret.) USA
Former FEMA Senior Official
Academe & Policy Research Senior Advisory Committee
Ex-Officio: Judge William Webster
Partner, Milbank, Tweed, Hadley & McCloy, LLP
Homeland Security Advisory Council Chair
Ex-Officio: Dr. James Schlesinger
Chairman, Board of Trustees, The MITRE Corporation
Homeland Security Advisory Council Vice Chair

Homeland Security Advisory Council Staff

Executive Director
Jeffrey D. Stern
Essential Technology Task Force
Charles J. Adams, Director
Amanda R. Rittenhouse, Editor
(DHS Honors Fellow)

Staff Directors
Candace Stoltz

Mike Miron

Staff Support
Jennifer Myers

APPENDIX B: ESSENTIAL TECHNOLOGY TASK FORCE SUBJECT MATTER EXPERTS

Subject Matter Experts

Dr. Penrose C. Albright
Managing Director, Civitas Group LLC
Jeff Barr
Senior Web Services Evangelist, Amazon Web Services
James E. Bennett
President and Chief Executive Officer, Metro Washington Airports Authority
Rich Beutel
Minority Staff, Senate Homeland Security & Government Affairs Committee
Christopher Bidwell
Managing Director Security, Air Transportation Association
Paul Bize
Vice President, Homeland Security and Intelligence for EDS U.S. Government Solutions
Scott Boylan
General Counsel, General Electric Company
Charles R. Chambers, Jr.
Senior Vice President, Security and Facilitation, Airports Council International-North America
Scott Charbo
DHS, Chief Information Officer
Patrick Ciganer
Executive Vice President for Customer Relations and External Affairs, In-Q-Tel
Jonah Czerwinski
Senior Fellow, Homeland Security, IBM Global Leadership Initiative

David Drabkin
Minority Staff, Senate Homeland Security & Government Affairs Committee
Elaine Duke
DHS, Acting Under Secretary for Management and (former) Chief Procurement Officer
Tom Essig
DHS, Chief Procurement Officer
Frank Finelli
Managing Director, The Carlyle Group
Mike Gaffney
President, Federal Sector Business Development, Computer Sciences Corporation
W. Scott Gould
Vice President, Public Sector Strategy and Growth, IBM Global Business Services
Claire Gradey
U.S. Coast Guard, Head of Contracting Activity
Jeff Green
Majority Staff, House Homeland Security Committee
John Hart
Financial Analyst, EDS
Robert Hooks
DHS, Director of Transition, Science & Technology Directorate
Dwight Hutchins
Global Managing Director, Accenture
Frank Inserra
DHS, Attorney Advisor for Office of General Counsel
Tim Malishenko
Vice President, Contracts and Pricing, The Boeing Company
Jack Mayer
Partner, Booz Allen Hamilton
Paul McNeill
Director, Homeland Security Programs, The Boeing Company
Steve Mertens
Chief, Homeland Security Branch, Executive Office of the President, Office of Management and Budget
Hans Miller
TSA Contractor, Hans Miller Strategy LLC

Daniel A. Rowley
General Counsel, GE Enterprise Solutions
Michael D. Russell
DHS, Deputy Associate General Counsel for General Law & Appropriations Counsel
Peter Sand
DHS, Director of Privacy Technology
Amelia Shachoy
Assistant Director, Government Accountability Office
Victoria Smith
Deputy Program Manager, DHS CIO HQ Support, MITRE
Michael Tangora
U.S. Coast Guard, Deputy Assistant Commandant for Acquisition and Director of Acquisition Services
Adam Tsao
Chief of Staff, Operational Processes and Technology, TSA
Bruce Walker
Vice President Strategic Planning – Homeland Security, Northrop Grumman

References

[1] Protect Our Nation from Dangerous People, 2) Protect Our Nation from Dangerous Goods, 3) Protect Critical Infrastructure, 4) Build a Nimble, Effective Emergency Response System and a Culture of Preparedness, and 5) Strengthen and Unify DHS Operations and Management.

[2] 1) Transportation Security Administration, 2) Customs and Border Protection, 3) Citizenship and Immigration Services, 4) Immigration and Customs Enforcement, 5) Federal Emergency Management Agency, 6) U.S. Secret Service, and 7) U.S. Coast Guard.

[3] IBM Global Business Services (Gould, Prieto, Czerwinski). *Global Movement Management; Strengthening Commerce, Security and Resiliency* in Today's Networked World. 2007. http://www-935.ibm.com/services/us/index.wss/ executivebrief/gbs/a1028853?cntxt=a1000055

[4] Department of Defense Office of the Under Secretary of Defense for Acquisition, Technology and Logistics. Strategic Goals Implementation Plan. 2008. http://www.acq.osd.mil/goals/20080207_SGIP.pdf

[5] Homeland Security Institute. Venture Capital Concept Analysis. 2005. www.homelandsecurity.org

In: Maintaining Homeland Security
Editors: A. P. Proctor et al., pp. 79-112
ISBN: 978-1-60692-990-2

Chapter 4

INFORMATION TECHNOLOGY: MANAGEMENT IMPROVEMENTS NEEDED ON THE DEPARTMENT OF HOMELAND SECURITY'S NEXT GENERATION INFORMATION SHARING SYSTEM*

United States Government Accountability Office

WHAT GAO FOUND

DHS halted further improvements on the existing HSIN system in September 2007. Since then, the department has continued to operate and maintain the system while a replacement—HSIN Next Gen—is being planned and acquired. DHS decided in large part to pursue this replacement due to

- the existing system has security and information-sharing limitations that do not meet department and other users' needs, thus impeding the department's ability to effectively perform its mission; and
- the new system is to be a key part of a departmentwide consolidation effort to, among other things, reduce the number of systems within DHS that share sensitive but unclassified information.

---

* This is an edited, excerpted and augmented edition of a U.S. Government Accountability Office Report, GAO-09-40, dated October 2008.

DHS has developed an acquisition strategy for HSIN Next Gen, whereby the system is to be implemented in four phases, each providing for an increasing number of users to be transitioned to the system. For example, DHS plans to begin transitioning existing HSIN users beginning in May 2009. Further, in May 2008, DHS issued a task order engaging a contractor to acquire, deploy, operate, and maintain the new system. The total estimated value of the task order's initial year is $19 million; the order also includes 4 option years that if exercised, are estimated to be worth $62 million. DHS intends to continue to use the existing HSIN with the goal of terminating its use in September 2009 when HSIN Next Gen is to be fully completed. DHS estimates it will cost $3.1 million to operate and maintain HSIN between now and its planned September 2009 termination.

DHS is in the process of implementing key acquisition management controls for HSIN Next Gen, but has yet to implement the full set of controls essential to effectively managing information technology system projects in a rigorous and disciplined manner. Specifically, it has not fully implemented key process controls in the areas of

- project and acquisition planning,
- requirements development and management, and
- risk management.

DHS officials, including the Office of Operations Coordination and Planning's Chief Information Officer, who is responsible for managing the project, attribute the partial implementation of these key processes in large part to the aggressive schedule for acquiring and deploying HSIN Next Gen. The Chief Information Officer also stated the department plans to address these weaknesses by, for example, tasking its contractor to assist in the development and completion of the risk management process area, but had not yet established dates for when all of these activities will be completed. Until these weaknesses are effectively addressed and DHS implements and institutionalizes the full set of acquisition management controls, the project will be at increased risk of operating in an ad hoc and chaotic manner— potentially resulting in increased project costs, delayed schedules, and performance shortfalls.

WHY GAO DID THIS STUDY

The Department of Homeland Security (DHS) is responsible for coordinating the federal government's homeland security communications with all levels of government. In support of this mission, DHS implemented, and has been enhancing, the Homeland Security Information Network (HSIN). It also has proposed a follow-on system, called Next Generation HSIN (HSIN Next Gen). GAO was asked to determine whether (1) DHS has stopped further improvements on HSIN and if so, the department's rationale for doing so and plans for acquiring its proposed follow-on system HSIN Next Gen and (2) the department is effectively managing the HSIN Next Gen acquisition. To accomplish this, GAO analyzed documentation, interviewed officials, and compared acquisition management processes and practices defined in industry best practices with those planned and underway by DHS.

WHAT GAO RECOMMENDS

GAO recommends strengthening acquisition management controls before the department starts to migrate existing users to the new system by, among other things, staffing the program office appropriately, ensuring all user requirements are gathered, and identifying key risks surrounding the project. In written comments on this chapter, DHS described actions planned and underway to address GAO recommendations.

ABBREVIATIONS

DHS	Department of Homeland Security
CIO	Chief Information Officer
HSIN	Homeland Security Information Network
HSIN	Next Gen Next Generation HSIN
IT	information technology
OPS	Office of Operations Coordination and Planning

October 8, 2008
The Honorable Joseph I. Lieberman
Chairman
The Honorable Susan M. Collins
Ranking Member
Committee on Homeland Security and Governmental Affairs
United States Senate

The Honorable Bennie G. Thompson
Chairman
Committee on Homeland Security House of Representatives

The Department of Homeland Security (DHS) is responsible for coordinating the federal government's homeland security communications with all levels of government—including state and local. In support of this mission, the department deployed, and has been making improvements to, the Homeland Security Information Network (HSIN) as part of its goal to establish an infrastructure for sharing homeland security information. In 2005,[1] and more recently in January 2007,[2] we designated homeland security information sharing as a high-risk area. Consequently, it is important that federal networks and associated systems, applications, and data facilitate this vital information sharing, and do so in a manner that produces effective information sharing among and between the various levels of government. This is particularly crucial for DHS's HSIN, which is the department's primary information technology (IT) system for sharing terrorism and related information. Recently, DHS proposed a follow-on system to HSIN, which it refers to as Next Generation HSIN (HSIN Next Gen).

This chapter responds to your request that we determine whether (1) DHS has stopped further improvements on HSIN and if so, the department's rationale for doing so and plans for acquiring its proposed follow-on system HSIN Next Gen and (2) the department is effectively managing the HSIN Next Gen acquisition.

On July 11, 2008, and July 17, 2008, we provided a briefing to staff of the House Homeland Security Committee and Senate Homeland Security and Governmental Affairs Committee, respectively. Prior to these staff briefings (on July 9, 2008), we provided the briefing to responsible DHS officials, who agreed in large part with our findings, conclusions, and recommendations. This chapter transmits (1) the slides that we used during the briefings and (2) the recommendations that we made to the Secretary of Homeland Security and the Director, Office of Operations Coordination and Planning, who is responsible for managing HSIN and HSIN Next Gen. The full briefing, including our scope and methodology, is reprinted as appendix I.

DHS Has Stopped Current HSIN System Improvements and Is in the Process of Acquiring a Replacement System

In September 2007, the department halted further improvements on the existing HSIN system. Since then, DHS has continued to operate and maintain the system while its replacement—HSIN Next Gen—is being planned and acquired. The department decided to pursue the replacement for two reasons. First, the existing system has security and information- sharing limitations that do not meet department and other users' needs. For example, with regard to security, the current HSIN does not support role-based access controls[3] and two-factor authentication.[4] These limitations hinder the department's ability to effectively perform its mission.

Second, the replacement system is to be used as a key part of a departmentwide consolidation effort aimed at reducing the number of multiple portals or Web-based systems within DHS by consolidating the systems across the department that are to share sensitive but unclassified information. In particular, HSIN Next Gen is to provide secure access to DHS sensitive but unclassified information and services for all department user communities, including those in the law enforcement, intelligence, immigration, and emergency and disaster management communities.

With regard to DHS plans to acquire HSIN Next Gen, the department has developed an acquisition strategy for the system and plans to have all users on the new system by September 2009. The system will be implemented in four phases, each addressing a functional portion of the requirements and providing for an increasing number of users to be transitioned to the system. Specifically, during the first phase of implementation, the department plans to bring on board up to 20,000 new users from critical infrastructure sectors such as agriculture and food, and transportation systems. In addition, during the second phase (called Initial Operational Capability) and third phase (called Maturing Operational Capability), DHS plans to transition over 26,000 users that currently use the existing HSIN system; this transition of existing HSIN users is to begin in May 2009. To help carry out the strategy, DHS issued a task order in May 2008 engaging a contractor to acquire, deploy, operate, and maintain the new system. The total estimated value of the base year of this arrangement is $19 million, and the total estimated value, if each of the four options is exercised, is $62 million.

DHS intends to continue to use the existing HSIN with the goal of terminating its use in September 2009 when HSIN Next Gen is to be fully implemented. DHS estimates it will cost $3.13 million to operate and maintain HSIN between now and its planned September 2009 termination.

DHS Has Yet to Implement the Management Controls Essential to Effectively Manage the HSIN Next Gen Acquisition

As we have previously reported,[5] the success of critical projects such as HSIN depends on developing and implementing a full set of acquisition management controls to effectively manage the project. While DHS is in the process of implementing key acquisition management controls for HSIN Next Gen, it has yet to implement the full set of controls essential to managing HSIN Next Gen in a disciplined and rigorous manner. Specifically, it has not implemented key process controls in the areas of

- project and acquisition planning, which includes key processes, such as developing a program office and identifying staff roles and responsibilities;
- requirements development and management, which involves key processes, such as gathering, analyzing, and validating user requirements; and
- risk management, which includes key processes, such as identifying and analyzing risks and assigning responsibilities for managing risks.

With regard to project and acquisition planning, DHS has established a program office for HSIN Next Gen, including filling the position of project manager. However, it has not adequately staffed the HSIN Next Gen program office and identified staff roles and responsibilities.

In addition, in the area of requirements development and management, the department has gathered and analyzed requirements from critical infrastructure sector users. However, it has not gathered requirements from all other HSIN users and developed a change control process for managing change to requirements.

Further, regarding risk management, DHS has begun to develop a risk management plan that defines staff roles and responsibilities. However, it has yet to identify all key risks surrounding the project and develop risk mitigation plans and completion milestones.

DHS officials, including the Office of Operations Coordination and Planning's (OPS) Chief Information Officer (CIO), who is responsible for managing the project, attribute the partial implementation of these key processes in large part to the aggressive schedule for acquiring and deploying HSIN Next Gen. In our view, engaging a contractor and commencing work before implementing mature controls is not a recipe for success. Specifically, our research and experience at federal agencies have shown that the probability of success is low using this approach. The OPS CIO stated the department plans to address these weaknesses by, for example, tasking its contractor to assist in the development and completion of the risk management process area, but had not yet established dates for when all of these activities will be completed.

Consequently, until these weaknesses are effectively addressed and DHS implements and institutionalizes the full set of acquisition management controls, the project will be at increased risk of operating in an ad hoc and chaotic manner—potentially resulting in increased project costs, delayed schedules, and performance shortfalls.

CONCLUSIONS

DHS has been challenged in its ability to efficiently and effectively manage the department's existing primary information-sharing system. In particular, although DHS has invested upwards of $70 million on the system, it still does not fully meet user needs and as a result, has not been fully utilized. DHS intends to address this performance shortfall by, among other things, acquiring a replacement system. A key challenge for DHS in this effort will be ensuring it develops an information-sharing system that effectively addresses its users' needs and in the process, does not waste or unwisely invest critical department resources.

To its credit, DHS has initiated some important steps in establishing sound and capable acquisition controls, but much remains to be accomplished before DHS management efforts can be considered effective and thereby minimize the risks associated with HSIN Next Gen delivering promised capabilities and benefits on time and within budget.

Investing money given the current state of management controls puts the project at risk. Given what is at stake, it is extremely important that DHS direct its attention to these management issues, and mitigate the associated risks as soon as possible.

RECOMMENDATIONS FOR EXECUTIVE ACTION

To minimize risks to the HSIN Next Gen project, we are making six recommendations to the Secretary of Homeland Security aimed at strengthening management of the project. We recommend that the Secretary direct the Director, Office of Operations Coordination and Planning to strengthen program management controls by

- staffing the program office appropriately;

- identifying staff roles and responsibilities;
- ensuring all requirements are gathered, analyzed, and validated;
- developing and implementing a requirements change control process; and
- ensuring effective risk management by identifying all key risks surrounding the project and developing risk mitigation plans and completion milestones.

We also recommend that these controls be implemented before the department starts to migrate users to HSIN Next Gen's Initial Operational Capability.

Agency Comments and Our Evaluation

In written comments on a draft of this chapter, which were in a letter signed by DHS's Director of Operations Coordination and Planning and are reprinted in appendix II, the department described actions planned and underway to address our recommendations. These actions are consistent with those described by DHS in response to our July 9, 2008, briefing to the department in which it largely agreed with our findings, conclusions, and recommendations.

David A. Powner
Director, Information Technology Management Issues

Appendix I: Briefing Slides to Congressional Staff

Information Technology: Management Improvements Needed on the Department of Homeland Security's Next Generation Information Sharing System

Briefing to the Staffs of the
Senate Committee on Homeland Security and Governmental Affairs July 17, 2008
House Committee on Homeland Security July 11, 2008

- Introduction
- Objectives, Scope, and Methodology

INTRODUCTION

The Department of Homeland Security (DHS) is responsible for coordinating the federal government's homeland security communications with all levels of government—including state and local. In support of this mission, the department implemented, and has been enhancing, the Homeland Security Information Network (HSIN) as part of its goal to establish an infrastructure for sharing homeland security information.[6] Recently, DHS proposed a follow-on system to HSIN, which it refers to as Next Generation HSIN (HSIN Next Gen).

In 2005,[7] and more recently in January 2007,[8] we designated homeland security information sharing as a high-risk area. Consequently, it is important that federal networks and associated systems, applications, and data facilitate this vital information sharing, and do so in a manner that produces effective information sharing among and between the various levels of government. This is particularly crucial for DHS's HSIN, which is the department's primary information technology (IT) system for sharing terrorism and related information.

OBJECTIVES, SCOPE, AND METHODOLOGY

As agreed, our objectives were to determine whether

- DHS has stopped further improvements on HSIN and if so, the department's rationale for doing so and plans for acquiring its proposed follow-on system called HSIN Next Gen system, and

- the department is effectively managing the HSIN Next Gen acquisition.

For our first objective, we analyzed documentation and interviewed DHS officials from the office responsible for managing HSIN and HSIN Next Gen, the Office of Operations Coordination and Planning (OPS), to assess efforts planned and underway to implement HSIN system improvements and acquire HSIN Next Gen.

For our second objective, we compared processes and practices defined in the Software Engineering Institute's Capability Maturity Model® Integration for Acquisition (CMMIACQ)[9] and in our prior work analyzing best practices in industry and government[10] with those planned and underway by the department to determine the extent of implementation. In judging implementation, we used the following criteria: the processes were (1) *fully implemented* if all of the related guidance was addressed; (2) *partially implemented* if some, but not all, of the related guidance was addressed; and (3) *not implemented* if none of the related guidance was addressed.

Details of our scope and methodology are provided in attachment I. We conducted this performance audit from January 2008 to June 2008, in accordance with generally accepted government auditing standards. Those standards require that we plan and perform the audit to obtain sufficient, appropriate evidence to provide a reasonable basis for our findings and conclusions based on our audit objectives. We believe that the evidence obtained provides a reasonable basis for our findings and conclusions based on our audit objectives.

Results in Brief

DHS halted further improvements on the existing HSIN system in September 2007. Since then, the department has continued to operate and maintain the system while a replacement—called HSIN Next Gen—is being planned and acquired. DHS decided to pursue this replacement for two reasons.

- First, the existing system has security and information sharing limitations that do not meet department and other users' needs, thus impeding the department's ability to effectively perform its mission.
- Second, the replacement system is to be used as a key part of a departmentwide consolidation effort to reduce the number of duplicative DHS Web-based systems.

DHS has developed an acquisition strategy for the system and plans to have all users on the new system by September 2009. DHS intends to continue to use the existing HSIN with the goal of retiring it once HSIN Next Gen has been completed. DHS estimates it will cost $3.13 million to operate and maintain HSIN between now and retirement.

DHS is in the process of implementing key acquisition management controls for HSIN Next Gen. For example, DHS has established a program office for HSIN Next Gen, including filling the position of project manager. In addition, DHS has begun to develop a risk management plan that defines staff roles and responsibilities. However, DHS has yet to implement the full set of controls essential to effectively managing IT system projects in a rigorous and disciplined manner. Specifically, it has not fully implemented key process controls in the areas of

- project and acquisition planning,
- requirements development and management, and
- risk management

DHS officials, including the OPS Chief Information Officer (CIO), who is responsible for managing the project, attribute the partial implementation of these key processes in large part to the aggressive schedule for acquiring and deploying HSIN Next Gen. DHS has activities planned and underway to address missing controls, but has not established dates for when all of these activities will be completed.

Until DHS has implemented these controls, there is increased risk of the project operating in an ad hoc and chaotic manner—potentially resulting in increased project costs, delayed schedules, and performance shortfalls. Accordingly, we are making recommendations to the Secretary of Homeland Security to (1) strengthen management controls, including project and acquisition planning, requirements development and management, and risk management; and (2) ensure that these controls be implemented before users are transitioned to HSIN Next Gen Initial Operational Capability.

In orally commenting on a draft of this briefing, DHS officials stated that they agreed with our findings and recommendations and described actions they have initiated to implement our recommendations. They also generally agreed with our conclusions. However, DHS officials stated that the risk raised in our conclusions was mitigated by their IT experience. While experience is important, key process controls, such as rigorous and disciplined requirements and risk management, are also essential to IT project success.

BACKGROUND

DHS is the lead department involved in securing our nation's homeland. Its mission includes, among other things, leading the unified national effort to secure the United States, preventing and deterring terrorist attacks, and protecting against and responding to threats and hazards to the nation.

As part of its mission and as required by the Homeland Security Act of 2002,[11] the department is also responsible for coordinating efforts across all levels of government and throughout the nation, including with federal, state, tribal, local, and private sector homeland security resources. This includes coordinating the federal government's networks and other communications systems with state and local governments.

In 2004, DHS developed and implemented HSIN as the department's primary IT system for sharing terrorism and related information with federal, state, and local agencies, among others. Specifically, this Web-based communication system is to provide a secure and trusted national IT system for sensitive but unclassified information sharing and collaboration among federal, state, local, tribal, territorial, private sector, and international partners engaged in preventing, protecting from, responding to, and recovering from all threats, hazards, and incidents within DHS's authority.

HSIN offers both real-time chat and instant messaging capability, as well as a document library that contains reports from multiple federal, state, and local sources. Available through the system are suspicious incident and pre-incident information and analysis of terrorist threats, tactics, and weapons. Each community of interest has Web pages that are tailored for the community and contain general and community-specific news articles, links, and contact information.

HSIN is to support a number of homeland security-related mission areas that cover thousands of users across the United States. These mission areas include over 35 user groups, commonly referred to as communities of interest, including

- emergency management,
- law enforcement,
- counterterrorism,
- individual states, and
- private sector communities.

Other DHS component organizations, such as the Office of Infrastructure Protection, the Coast Guard, and Federal Emergency Management Agency, use HSIN as a tool to further their respective missions and therefore have assisted in the development, operations and maintenance, and enhancement of HSIN. For example, according to the Office of Infrastructure Protection, it works with the critical infrastructure sectors—that is, groups of similar private and government entities that operate and maintain systems and assets, whether physical or virtual, so vital to the nation that their incapacity or destruction would have a debilitating impact on national security, national economic security, national public health or safety, or any combination of those matters[12]—to gather user requirements and develop business processes in order to integrate HSIN into the critical sectors' information-sharing environment.

The Office of Operations Coordination and Planning (OPS) CIO is responsible for ensuring that HSIN supports the needs of the department and its partners. This includes managing HSIN operations and maintenance, making necessary enhancements to the current system, and developing and acquiring HSIN Next Gen. The OPS CIO reports directly to the OPS Director who in turn reports directly to the DHS Secretary and Deputy Secretary.

Through fiscal year 2007, the department reports it has expended about $70 million on HSIN, and for fiscal year 2008, the department had budgeted about $21 million for operations, maintenance, and enhancement.

In April 2007,[13] we reported that when coordinating efforts between HSIN and other state and local information-sharing initiatives, DHS did not fully adhere to key practices aimed at enhancing information sharing, collaboration, and avoiding duplication. For example, in developing the system, the department did not work with two key state and local initiatives, which are major parts of the Regional Information Sharing System program, to fully develop joint strategies to meet mutual needs.

In addition, it did not develop compatible policies, procedures, and other means to operate across organizational boundaries. DHS's limited use of these practices was attributable to a number of factors, including the department's expediting its schedule to deploy information-sharing capabilities after the events of September 11, 2001, and in doing so not developing a comprehensive inventory of key state and local information- sharing initiatives.

As a result, we found there was increased risk that, among other things, effective information sharing is not occurring. Additionally, the department risked duplication of state and local capabilities. We recommended, among other things, that DHS

- identify and develop a comprehensive inventory of state and local initiatives;

- assess whether there are opportunities for HSIN to improve information sharing and avoid duplication of effort; and
- where there are opportunities, implement effective coordination and collaboration practices.

In response, DHS largely agreed with our recommendations and initiated actions to implement them. Examples include the following:

- In October 2007 and in February 2008, the HSIN Advisory Council—a HSIN user group composed of representatives from state, tribal, and local governments and the private sector—met to discuss HSIN information-sharing activities and provided strategic-level recommendations to the OPS Director.
- The HSIN Mission Coordinating Committee—a user group composed of representatives from DHS's components (e.g., the Office of Infrastructure Protection, the Coast Guard, and the Federal Emergency Management Agency)—has met five times over the past year to address their respective users' requirements for HSIN.

In July 2007,[14] we reported on challenges the department faced when using HSIN to share information with critical infrastructure sectors. Examples included:

- DHS officials responsible for leading the national effort to reduce critical infrastructure risk stated that although they encouraged critical sector entities to use HSIN, the system did not provide the capabilities that were promised, including providing the level of security expected by certain sectors.
- An internal DHS review of HSIN determined that the department had not clearly defined the purpose and scope of the system, and that the system had been developed without sufficient planning and project management.

RESULTS: OBJECTIVE 1

HSIN Is Currently Operational but Further Improvements Have Been Halted

DHS Has Stopped Current HSIN System Improvements and Is in the Process of Acquiring a Replacement System

The department halted further HSIN improvements in September 2007 but it continues to operate and maintain the system while its replacement—HSIN Next Gen—is being planned and acquired.

DHS decided to pursue a replacement system based on two reasons. First, the current system has security and information-sharing limitations that do not meet its users' needs and thus impedes the department's ability to effectively perform its mission. Second, the new system is to be used as part of a departmentwide effort—referred to as the portal consolidation program—to consolidate multiple portals or Web-based systems and improve sensitive but unclassified information-sharing capabilities within the department.

DHS has developed a HSIN Next Gen acquisition strategy and as part of the strategy, issued a May 2008 task order engaging a contractor to develop the system. DHS plans to have all users on the new system by September 2009. In the interim, DHS plans to continue to operate and maintain HSIN as the new system is acquired and deployed and users are transitioned to it. Once user transition is complete, the department intends to retire HSIN.

In September 2007, DHS executives, including the Undersecretary for Management, Chief Information Officer, Director of Operations Coordination and Planning, and key system user representatives (e.g., Office of Infrastructure Protection), met to discuss HSIN operations. Key representatives said HSIN was not meeting their needs due to system security and information-sharing limitations.

System security limitations cited included the system's inability to support

- role-based access controls, which limit system functions based on a user's designated role, and
- two-factor authentication, which is a way of verifying someone's identity by using two of the following: something the user knows (password), something the user has (badge), or something unique to the user (fingerprint).

Information-sharing limitations included the system's inability to

- enable users to access HSIN and systems outside of DHS (such as the state and local law enforcement's Regional Information Sharing System) using single sign capability (i.e., requiring only one user name and password);
- enable users to send alerts and notifications and receive alerts through e-mail or cell phones;
- support online meetings and presentations; and
- upload new users into the system in bulk.

HSIN Improvements Halted

According to user representatives, these limitations were hindering their ability to perform the mission of the department. For example, representatives from the Office of Infrastructure Protection (which is part of the National Protection and Programs Directorate) stated that without the security controls, private-sector officials from the critical infrastructure sectors were reluctant to share with DHS sensitive information about sector infrastructure that is essential to protecting the homeland, thus inhibiting the department's ability to adequately build trusted relationships with sector officials. In response, the Office of Infrastructure Protection initiated an effort to obtain requirements from HSIN critical infrastructure sectors users, augmenting the requirements the department had for the existing system.

Consequently, the executives at the September 2007 meeting (referenced above) decided the best way to implement the missing security and information-sharing capabilities was via a new system, rather than by enhancing the existing system. According to these officials, they based their decision largely on the view that the existing system could not be enhanced to provide these capabilities in a cost-effective manner. These officials also decided at this time to halt any further HSIN enhancements until the new system (HSIN Next Gen) was implemented, at which point they planned to retire the current HSIN system.

HSIN Next Gen's Goal Is to Also Eliminate Duplication

In addition, in October 2007 the Under Secretary for Management issued a memorandum detailing how HSIN Next Gen is to be used as an integral part of the department's portal consolidation program. According to the memorandum, the current DHS Web environment consists of more than 100 Web-based systems, which are mostly duplicative in capabilities. HSIN Next Gen is part of a departmentwide program aimed at reducing the number of duplicative Web-based systems within DHS by consolidating the systems across the department that are used to share sensitive but unclassified information, and by replacing portal technologies that limit its information-sharing capabilities.

In particular, according to the memorandum, HSIN Next Gen is to provide secure access to DHS information and services for all DHS user communities, including those in the law enforcement, intelligence, immigration, and emergency and disaster management communities.

Homeland Security Information Network Next Generation

As part of the system acquisition and implementation strategy, DHS plans to continue operating and maintaining HSIN until September 2009. The department estimates the cost to operate and maintain the current system through September 2009 will be $3.13 million. DHS reports it will have spent a total of $91 million on HSIN by the end of fiscal year 2008.

In parallel, the department plans to begin developing and implementing HSIN Next Gen in four phases; the phases—along with a brief description of their functional purpose—are as follows.

- Phase one, referred to as Spiral 1, is to establish an operational platform for the HSIN critical sector users' requirements.
- The second phase, Initial Operational Capability, is to (1) deliver requirements currently supported by HSIN, as well as provide additional security controls and (2) begin migrating users of the current system to HSIN Next Gen.
- Phase three, Maturing Operational Capability, is to migrate all remaining users of the current system to HSIN Next Gen.
- The fourth phase, called the Final Operational Capability, is to provide for improved content management; better information discovery and delivery; and improved alert, notification, and public announcement functions.

Each phase is intended to, among other things, address a functional portion of the requirements and provide for an increasing number of users to be transitioned to the system. In addition, DHS plans to draw upon the existing HSIN system and capabilities, rather than developing a complete infrastructure replacement. Specifically, where possible, it plans to re-use existing HSIN hardware and software. The department plans to use the contractor (discussed in detail below) to help them do this. However, it has yet to set a date for when this is to be completed.

Further, in terms of users, during the first phase of implementation, the department plans to bring on board up to 20,000 critical sector users. In addition, over the second and third phases, DHS plans to transition over 26,000 users that currently use the existing HSIN system.

In May 2008, the department issued a task order to a contractor[15] to acquire, deploy, operate, and maintain the new system. The total estimated value of the base year of this arrangement is $19 million, and the total estimated value, if each of the four options is exercised, is $62 million.

Each of the HSIN Next Gen phases, the timing of their implementation, the percentage of users to be transitioned, and the date the contractor was issued the task order are depicted in figure 1.

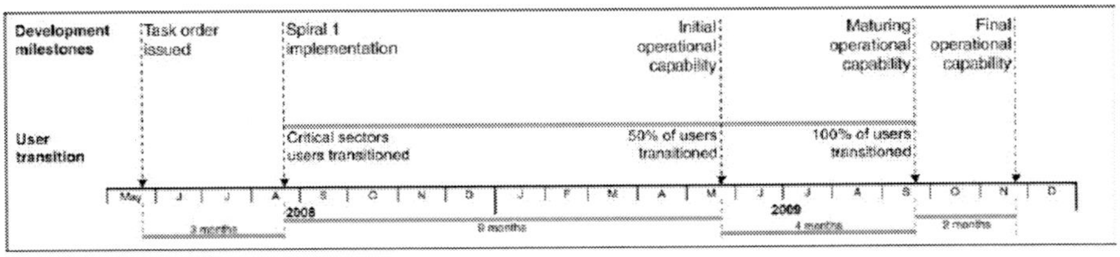

Source: GAO analysis of DHS data.

Figure 1. HSIN Next Generation Phases and Associated Milestones.

Key dates are:

- May 2008 – issued task order to contractor for HSIN Next Gen.
- August 2008 – implement Spiral 1 with the goal of supporting up to 20,000 critical sectors users.
- May 2009 – complete Initial Operational Capability with 13,000 current users scheduled to transition.
- September 2009 – implement Maturing Operational Capability with the transition of the remaining 13,000 users.
- November 2009 – complete Final Operational Capability by delivering new functionality to users.

Results: Objective 2

Acquisition Management Controls Needed

DHS Has Yet to Implement the Management Controls Essential to Effectively Manage the HSIN Next Gen Acquisition

DHS is in the process of implementing key acquisition management controls, but it has yet to implement the full set of controls essential to managing HSIN Next Gen in a disciplined and rigorous manner. Specifically, it has not implemented key process controls in the areas of

- project and acquisition planning,
- requirements development and management, and
- risk management.

Until DHS has fully implemented these controls, it increases the risk of the project operating in an ad hoc and chaotic manner—potentially resulting in increased project costs, delayed schedules, and performance shortfalls.

As we have previously reported,[16] the success of critical projects such as HSIN depends on developing and implementing a full set of acquisition management controls to effectively manage the project. Leading organizations, such as the Software Engineering Institute and the Chief Information Officer's Council, and our research and experience at federal agencies have shown that such process controls are significant in successful system acquisition and development projects. In particular, the CMMI-ACQ[17] has defined a suite of key acquisition process control areas that are necessary to manage system acquisitions in a rigorous and disciplined fashion. These process areas include

- project and acquisition planning,
- requirements development and management, and
- risk management.

The following table provides a list of key processes within each process area.

DHS is currently implementing key acquisition controls for the HSIN Next Gen but it has yet to implement the full set of controls essential to effectively managing the project.

Table 2 provides a summary of the status of the project relative to each of the key process areas.

Table 1. Key Processes for Effectively Managing IT Projects

Project and acquisition planning	• developing a program office • obtaining appropriate staff, and ensuring that staff have the skills and knowledge needed to manage the project • identifying staff roles and responsibilities • identifying key deliverables and milestones for the project and acquisition
Requirements development and management	• gathering user requirements • analyzing and validating user requirements • managing any changes to the requirements in collaboration with stakeholders
Risk management	• identifying and analyzing risks • assigning responsibilities for managing risks • developing mitigation plans and completion milestones for identified risks

Source: GAO summary of leading practices, including practices identified by the Software Engineering Institute, the Chief Information Officer's Council, and the Office of Management and Budget.

Table 2. Summary of the Status of HSIN Next Gen Acquisition Management Controls as of June 2008

Process area	Key processes	Status
Project and acquisition planning	• Establish a program office • Obtain appropriate staff • Identify staff roles and responsibilities • Identify key deliverables and milestones for project and acquisition	● ○ ○ ●
Requirement development and management	• Gather user requirements • Analyze and validate user requirements • Manage change to requirements	◘ ◘ ○
Risk management	• Identify and analyze risks • Assign responsibilities for managing risks	◘ ● ○

	• Develop mitigation plans and completion milestones for identified risks	

Source: GAO analysis of agency data.
● Key process area implemented
◘ Key process area partially implemented
○ Key process area not implemented.

With regard to project and acquisition planning, DHS has implemented two of the four key processes. Specifically, it has

- established a program office for HSIN Next Gen, including filling the position of the project manager, and developed an April 2008 mission needs statement for HSIN Next Gen; and
- developed a project schedule, identifying key deliverables and milestones, for the HSIN Next Gen project and acquisition.

However, having already issued a task order to the contractor for HSIN Next Gen, the department has not filled two positions that it identified it needed to appropriately staff the program office. According to DHS officials, including the OPS CIO, they are in the process of hiring two full-time employees by the end of fiscal year 2008. In addition, the department is in the process of identifying staff roles and responsibilities, but has yet to finalize the effort.

Until the program office is adequately staffed and roles and responsibilities have been defined, DHS will be challenged in its ability to manage the HSIN Next Gen acquisition and project, including overseeing the contractor tasked to develop the system.

With regard to requirements development and management, DHS has partially implemented two of the three key processes, and has yet to implement the remaining process. Specifically, for Spiral 1, DHS has

- gathered user requirements from the critical infrastructure sector users, and
- analyzed these requirements through the OPS CIO, HSIN stakeholders, and the HSIN Mission Coordinating Committee.

The department used these user requirements, the existing HSIN requirements, and pending change requests for the current system to create the Functional Requirements Document dated March 2008. This document defines and outlines the known user requirements for HSIN Next Gen. The Functional

Requirements Document was included as part of the HSIN Next Gen solicitation documentation (i.e., request for proposals) used to award the contractor in May 2008. However, while DHS has gathered and analyzed user requirements from critical infrastructure sector users, it has not gathered requirements from all other HSIN users. Moreover, DHS has yet to validate the requirements.

In addition, DHS has not developed a change control process for managing change to requirements in collaboration with stakeholders, including developing criteria for evaluation and acceptance of requirements.

DHS has efforts planned and underway to address these weaknesses. For example, the department is in the process of establishing an initiative (called the HSIN Mission Integration Effort) to improve its ability to gather user requirements by having a formal outreach process to communicate with HSIN users. According to the OPS CIO, this is part of the department's effort to improve its capability to gather requirements from HSIN users. In addition, DHS plans to validate requirements for each HSIN Next Gen phase before they are completed, which is to be by August 2008 for Spiral 1. Further, DHS plans to establish a change control board to manage HSIN Next Generation requirements by September 2008.

While these are steps in the right direction, until they are completed and DHS has fully gathered, analyzed, and validated all user requirements and implemented effective change management, it faces the risk that HSIN Next Gen will not meet user and mission needs, which is a problem it faced with the existing HSIN and why it is currently working on a replacement system.

With regard to risk management, DHS has implemented one of the key processes and part of another, and has yet to implement the remaining process. Specifically, DHS's HSIN Next Gen Acquisition Plan (dated February 2008)

- assigns responsibility for managing the risks; and
- partially identifies a list of primary risks both internal and external to the department, such as
    - insufficient funding to execute future development,
    - insufficient government staff to execute the project, and
    - changes in HSIN user requirements that could negatively impact cost and schedule.

In addition to these efforts, DHS has begun to develop a risk management plan that defines staff roles and responsibilities, including procedures for identifying and tracking risks and assessing the probability and impact of individual risks.

However, the department has yet to develop risk mitigation plans and completion milestones, which includes recommended courses of action for each critical risk. The department intends to develop such plans, which are to provide risk mitigation strategies with alternatives and mitigation project plans, including activities, schedules, and resource requirements. However, the department has yet to establish a date for when this is to be completed.

In addition, the list of primary risks prepared did not include all key risks. For example, HSIN Next Gen's schedule, which has been identified by the OPS CIO as being aggressive, has not been identified as a risk.

Until DHS fully implements and institutionalizes risk management, there is increased probability that unanticipated risks may occur that could have a critical impact on HSIN Next Gen's cost, schedule, and performance.

The OPS CIO stated that the reason for the partial implementation of these key processes is attributable in large part to an aggressive schedule for acquiring and deploying HSIN Next Gen.

In our view, engaging a contractor and commencing work before implementing mature controls is not a recipe for success. Specifically, our research and experience at federal agencies have shown that the probability of success is low using this approach. A case in fact is the existing HSIN system which was acquired and deployed via an overly aggressive schedule with the result being it did not meet all users' needs, necessitating in part the need for the HSIN Next Gen replacement.

The OPS CIO stated the department plans to address these weaknesses by, for example, tasking its contractor to assist in the development and completion of the risk management process area. However, until the processes have been implemented and institutionalized, and the full set of acquisition management controls are implemented, the project will be at increased risk of operating in an ad hoc and chaotic manner—potentially resulting in increased project costs, delayed schedules, and performance shortfalls.

Conclusions

DHS has been challenged in its ability to efficiently and effectively manage the department's existing primary information-sharing system. In particular, although DHS has invested upwards of $70 million on the system, it still does not fully meet user needs and as a result, has not been fully utilized. DHS intends to address this performance shortfall by, among other things, acquiring a replacement system. A key challenge for DHS in this effort will be ensuring it develops an information-sharing system that effectively addresses its users' needs and in the process, does not waste or unwisely invest critical department resources.

To its credit, DHS has initiated some important steps in establishing sound and capable acquisition controls, but much remains to be accomplished before DHS management efforts can be considered effective and thereby minimize the risks associated with HSIN Next Gen delivering promised capabilities and benefits on time and within budget.

Investing money given the current state of management controls puts the project at risk. Given what is at stake, it is extremely important that DHS direct its attention to these management issues, and mitigate the associated risks as soon as possible.

Recommendations for Executive Action

To minimize risks to the HSIN Next Gen project, we are making six recommendations to the Secretary of Homeland Security aimed at strengthening management of the project. We recommend that the Secretary direct the Director, Office of Operations Coordination and Planning to strengthen program management controls by

- staffing the program office appropriately;
- identifying staff roles and responsibilities;
- ensuring all requirements are gathered, analyzed, and validated;
- developing and implementing a requirements change control process; and
- ensuring effective risk management by identifying all key risks surrounding the project and developing risk mitigation plans and completion milestones.

We also recommend that these controls be implemented before the department starts to migrate users to HSIN Next Gen's Initial Operational Capability.

AGENCY COMMENTS AND OUR EVALUATION

In oral comments on a draft of this briefing, DHS officials agreed with our findings and recommendations and described actions that they have underway to address our recommendations. In particular, the OPS CIO stated that they have engaged a contractor to help them organize the HSIN program office, which includes identifying staff roles and responsibilities.

DHS officials also generally agreed with our conclusions. However, they took exception with the statement in our conclusions that investing money given the current state of management controls puts the project at risk. According to DHS officials, including the OPS CIO, they believe the risks to the project are mitigated by the IT experience of the HSIN staff, including the knowledge it has gained over the past 4 years in operating, maintaining, and enhancing HSIN. While we agree that IT experience is important, our research and experience at federal agencies have shown that, in addition to people, key processes, such as rigorous and disciplined requirements and risk management, are essential to IT project success.

DHS officials also provided technical comments, which we have incorporated into the briefing as appropriate.

ATTACHMENT I
SCOPE AND METHODOLOGY

To address our first objective, we

- assessed department efforts to stop HSIN system improvements by analyzing agency documentation and then discussing with agency officials via interviews. For example, we

  - reviewed executive-level correspondence, memos, strategies, and related documentation describing the department's plans for the current system, including ceasing system improvements and the reasons for doing this;
  - reviewed cost estimates to determine the planned costs of the operations and maintenance, and discussed the costs of enhancing the current system with OPS officials; and
  - interviewed OPS officials to clarify our understanding of the documentation and the department's rationale for choosing to develop the follow-on system.

- analyzed DHS plans for the proposed follow-on system. Specifically, we

  - evaluated the HSIN Next Gen acquisition plan, requirements document, request for proposals, and related documentation to determine what activities were planned and when they were to be accomplished; and
  - reviewed independent cost estimates to determine the planned costs for the development, operations, and maintenance of the new system.

To address our second objective, we assessed the extent to which the department was managing the acquisition of HSIN Next Gen based on the processes defined in the Software Engineering Institute's Capability Maturity Model® Integration for Acquisition (CMMI-ACQ).[18] In particular, we analyzed the department's efforts in acquisition planning, requirements development and management, and risk management. In doing so, we

- assessed HSIN Next Gen acquisition and project planning documentation and interviewed OPS officials to obtain key milestones;
- reviewed the HSIN Next Gen system requirements and interviewed officials from OPS and the Office of Infrastructure Protection, and representatives from HSIN governance bodies in order to understand how requirements were gathered and managed; and
- evaluated the HSIN Next Gen risks and risk management plan, and interviewed OPS officials to understand how risks were identified and are to be managed.

In making these judgments, we used the following criteria: processes were

- *fully implemented* if all of the related guidance was addressed;
- *partially implemented* if some, but not all, of the related guidance was addressed; and
- *not implemented* if none of the related guidance was addressed.

We conducted our work at DHS headquarters offices in Washington, D.C., and the Office of Infrastructure Protection in Arlington, Virginia. We conducted this performance audit from January 2008 to June 2008, in accordance with generally accepted government auditing standards. Those standards require that we plan and perform the audit to obtain sufficient, appropriate evidence to provide a reasonable basis for our findings and conclusions based on our audit objectives. We believe that the evidence obtained provides a reasonable basis for our findings and conclusions based on our audit objectives.

APPENDIX II: COMMENTS FROM THE DEPARTMENT OF HOMELAND SECURITY

Operations Coordination and Planning
U.S. Department of Homeland Security
Washington, DC 20528

Homeland Security

September 19, 2008

David A. Powner
Director, Information Technology Management Issues
U.S. Government Accountability Office

Dear Mr. Powner:

The Office of Operations Coordination and Planning (OPS) appreciates the opportunity to comment on the Government Accountability Office (GAO) report, "Information Technology: Management Improvements Needed on the Department of Homeland Security's Next Generation Information Sharing System." OPS in coordination with the DHS Office of the Chief Information Officer (OCIO) are working to establish a secure and trusted information sharing and collaboration environment for Sensitive but Unclassified (SBU) information for use by DHS and non-DHS partners engaged in preventing, protecting from, responding to, and recovering from all threats, hazards, and incidents within the authority of DHS.

The decision to upgrade the Homeland Security Information Network (HSIN) technology platform meets the growing needs of HSIN users. The current technology platform does not provide the necessary capabilities required to provide the necessary trust and interoperability. Upgrading HSIN technology addresses current user needs and provides a robust and trusted foundation adjustable over time to meet arising end user information sharing requirements. The project to upgrade the HSIN technology platform is called HSIN NextGen. It is important to understand that HSIN Next Generation (NextGen) is primarily a software upgrade to the current HSIN technology platform rather than an acquisition of a replacement system. The HSIN NextGen project, along with operations and maintenance of the current HSIN platform, is being done within the existing HSIN investment profile and does not require additional money.

The HSIN NextGen project will follow a phased implementation approach based on industry best practices. This approach allows the Department to effectively and efficiently move all current HSIN users onto an enhanced platform, constituting initial operating capability (IOC), by October, 2009. The current HSIN technology platform will remain operational throughout the phased implementation to ensure continued service for all users. Phase 1 of the HSIN NextGen project, driven by the HSIN Critical Sectors (HSIN-CS) priority requirements, was achieved on August 25, 2008.

The following responses address the recommendations within the report:

Recommendation: Staffing the program office appropriately

OPS has advertised for a HSIN Program Requirements Manager and is working with the DHS Office of the Chief Human Capital Officer to fill this position within 60 days. Two GS-15 leadership and technical positions have been advertised and we anticipate having personnel onboard within 60 to 90 days. The HSIN Program Manager is assisted by an experienced team of professional contracting firms. The roles filled by the contractors include cost, earned value management, schedule, performance, architecture, change process, and other support functions that are typical of a program management office.

The OPS CIO plans to fill ten additional billets beginning in Fiscal Year 2009 (FY09), pending Departmental approval. These billets will support architecture, security, privacy, and other functions. These specialists will ensure HSIN addresses statutory and interoperability requirements with partner tools. These resources will provide more robust requirements management and process control.

The HSIN program is not currently staffed to support simultaneous, significant outreach initiatives to our partners. To meet this demand, DHS plans to increase overall OPS CIO Division staffing in FY09 and FY10, subject to Congressional approval of existing budget requests. The augmented HSIN Outreach Team will build on our diverse partner community relationships to facilitate integrating HSIN into the partner communities' day-to-day operations that map to the DHS mission (Awareness, Prevent, Protect, Respond, and Recover). These new funds will be dedicated entirely to mission integration and focused on our Federal, State, local, and private sector partners.

Recommendation: Identifying staff roles and responsibilities

In April 2008, the OPS CIO initiated an effort by an outside team to analyze the current OPS CIO Division, which includes HSIN Program Management. This effort provided recommendations for transforming the OPS CIO Division and enable DHS to more effectively meet its complex, integrated mission requirements, both within DHS and across the larger homeland security community. The team conducted extensive research and performed over thirty-five interviews with OPS CIO staff, OPS stakeholders, and DHS-wide leaders. Then, the team applied proven analytical methods to form strategic and tactical views of organization, examining the CIO functions and capabilities it requires for the future. In addition, the team conducted a detailed, bottom-up assessment of existing capabilities and supporting activities. Four key areas were analyzed:

- **Process**: Develop an understanding of the existing and future processes including functions, tasks and activities needed to perform the mission of the OPS CIO Division
- **People**: Develop an understanding of the existing and future staff and expertise needed to support the mission and processes of the OPS CIO Division

- **Technology**: Develop an understanding of the existing and future technologies including applications, data and technology standards needed to perform the mission of the OPS CIO Division
- **Physical Infrastructure**: Develop an understanding of the existing, future facilities and working environment needed to perform the mission of the OPS CIO Division

The recommendation for the future state of the OPS CIO Division includes a detailed description of the organization model, including the processes, people, technology and infrastructure required, to implement the recommended organization.

Recommendation: Ensuring all requirements are gathered, analyzed, and validated

User requirements were the primary driver of the decision to upgrade the HSIN environment. These are not the only driver of the process. Initial phases will not meet every user requirement. The prioritization of certain user requirements is necessary. The Department must set timeline milestones in addition to identifying user requirements. This ensures that the awarded task order is completed in a timely manner, while initially ensuring that the Department meets the most urgent system requirements. Phases 1 through 3 of the HSIN NextGen project address the user needs to provide a secure and trusted information sharing and collaboration platform.

The Department determined that the HSIN NextGen project must first address the security and trust requirements identified through HSIN Community of Interest (COI) owners' input. Based upon input from many of the HSIN Community of Interest (COI) owners, the Department determined that the HSIN NextGen project must first address the security and trust requirements identified by all COIs. State, local and tribal first responders have reached out to the Department by requesting changes and sending requirements through the HSIN Helpdesk and/ or through the HSIN Mission Advocates. These change requests and requirements were recorded in the HSIN Change Request Tracking System (CHARTS). Many change requests were made by HSIN-CS and State, local, or tribal users. All change requests and requirements were examined and where possible incorporated into the HSIN NextGen Functional Requirements Document (FRD). The operational user requirements, which include policy, business process, and governance, will be gathered through identified DHS business leads and the HSIN Outreach Team.

Using a best practices approach, the HSIN Mission Integration Effort will gather user requirements and establish on-the-ground relationships through HSIN representatives (Mission Advocates). The HSIN Outreach Team is in the initial phase of an important engagement with the Commonwealth of Virginia, among others. Working closely with operational personnel in Virginia, the Department will further the understanding of the Commonwealth's information sharing needs and aid to support the Department partners' homeland security mission. In the future, the Department will engage with more partners to further examine the needs of our State, local, tribal and Federal partners.

The Department further determined that the most time sensitive and pressing needs of the existing HSIN COIs were those of the HSIN Critical Sectors (HSIN-CS). HSIN-CS provides a common environment for the critical infrastructure/key resource (CI/KR) stakeholder partners. NPPD has gathered and validated necessary user requirements for this phase from their stakeholders over a two year period. The critical infrastructure/key resource community is a well governed and defined community. The National Protection and Programs Directorate, Office of Infrastructure Protection (NPPD/IP), has determined that implementing the HSIN-CS priority requirements at the earliest moment was an absolute necessity to avoid mission degradation and loss of the voluntary participation of the 18 infrastructure sectors.

Recommendation: Developing and Implementing a requirements change control process

There must be one overarching requirements process that brings business, functional and technical architecture products into alignment. This is a complex undertaking, given the necessity for interoperability, as well as the depth, breadth, and volunteer nature of potential HSIN user groups. The phased approach to migrating communities onto the upgraded HSIN environment mitigates many risks.

The HSIN NextGen project will make the HSIN environment responsive and flexible to user requirements through a single, well-designed requirements process. The diversity of customer requirements and the need for a more standards-based platform, responsive to changing user requirements, is a driver for the HSIN NextGen project. The use of the maturing governance structure will ensure customer needs are met. The Information Sharing Governance Board (ISGB), along with the Information Sharing Coordinating Council (ISCC) and other mission coordination bodies, will work with the HSIN Program Manager to make certain that the requirements are captured, reviewed, and, if appropriate, implemented into the HSIN program change management process. DHS will adapt its tactics and timeline as needed using the phased deployment strategy and a segment architecture approach.

Future phases of HSIN NextGen will create improved versions based upon continued input from HSIN users. Currently, and moving forward in future phases, improvements to HSIN have been and will continue to be driven by the input of Federal, State, local, private sector, and tribal users with each phase improving upon the last. We anticipate that once HSIN users have a chance to understand and use the upgraded HSIN capabilities, they will suggest additional improvements or enhancements. These requests will translate into requirements to be submitted into the HSIN change management process and then incorporated into subsequent phases of the HSIN NextGen project.

To ensure success, a governance structure was initiated that integrated a larger segment architecture framework and the phased implementation approach. This structure continues to evolve to ensure that all stakeholders are involved and end user requirements are accurately captured, vetted, managed, and implemented. Key program activities and decisions are guided by DHS policies, processes, and procedures for consistency, repeatability, and compliance. The HSIN governance structure allows HSIN program resources to engage with mission leaders from all segments to determine whether HSIN is an appropriate solution for that target segment. If so, the governance structure allows us to identify mission requirements of that segment community and determine whether HSIN can meet those requirements in a timely, cost effective manner. The Department will move forward with the implementation of additional capabilities for new or existing mission areas based on whether HSIN can meet those requirements in a timely, cost effective manner. Once that determination is made, additional capabilities will be designed, developed, and validated with participation from stakeholders.

Recommendation: Ensuring effective risk management by identifying all key risks surrounding the project and developing risk mitigation plans and completion milestones

The HSIN Program Team exercises a proactive approach to risk. OPS identifies and mitigate risks before they manifest as schedule slippage, cost overruns, and unsatisfied requirements. Our risk management approach incorporates a continuing, closed-loop review and analysis of technical, programmatic, cost, and schedule risks throughout the entire program lifecycle. OPS uses proven management toolsets for detailed documentation and tracking of all identified risks/problems from point of discovery through risk resolution (e.g. web portals to facilitate user entry, tracking, reporting and maintenance of a centralized repository for all deliverables and product information). Our risk management approach monitors overall program health to ensure goals are being met. The Risk Management Plan consists of the following key areas:

- **Risk Identification**: Project managers are responsible for proactively identifying and documenting potential problems, issues, risks and dependencies at every program level
- **Risk Reporting**: Project managers conduct regular issue/risk review meetings to ensure risks are reported appropriately and in a timely manner. Prior to the internal program review (IPR), probability-of-occurrence and consequence-of-failure analyses are conducted to quantify and rank all identified risks
- **Risk Mitigation Strategy**: In addition to routine risk reporting, project manager are responsible for mitigation strategies for every risk that is identified. Impacted areas and/or systems, resources and skills required, as well as potential level of effort to provide resolution, are captured in the strategy
- **Risk Escalation**: Risks ranked high and medium may require special attention and/or action plans for mitigation, thus the overall Risk Management plans include an escalation path based on risk category and impacted area

I look forward to working with you to ensure that user communities that depend upon HSIN are able to accomplish their missions. If I may be of further assistance, please contact my office.

Sincerely,

Roger T. Rufe, Jr.

REFERENCES

[1] GAO, *High-Risk Series: An Update*, GAO-05-207 (Washington, D.C.: January 2005).

[2] GAO, *High-Risk Series: An Update*, GAO-07-310 (Washington, D.C.: January 2007).

[3] Role-based access controls limit system functions based on a user's designated role.

[4] Two-factor authentication is a way of verifying someone's identity by using two of the following: something the user knows (password), something the user has (badge), or something unique to the user (fingerprint).

[5] For example, GAO, *Information Technology: Management Improvements Needed on Immigration and Customs Enforcement's Infrastructure Modernization Program,* GAO-05-805 (Washington, D.C.: Sept. 7, 2005) and *Census Bureau: Important Activities for Improving Management of Key 2010 Decennial Acquisitions Remain to be Done,* GAO-06-444T (Washington, D.C.: Mar. 1, 2006).

[6] The Homeland Security Act of 2002 directed DHS to establish communications to share homeland security information with federal agencies, state and local governments, and other specified groups.

[7] GAO, *High-Risk Series: An Update,* GAO-05-207 (Washington, D.C.: January 2005).

[8] GAO, *High-Risk Series: An Update,* GAO-07-31 0 (Washington, D.C.: January 2007).

[9] Carnegie Mellon Software Engineering Institute, *Capability Maturity Model® Integration for Acquisition (CMMI-ACQ),* Version 1.2 (November 2007).

[10] For example, GAO, *Information Technology: Management Improvements Needed on Immigration and Customs Enforcement's Infrastructure Modernization Program,* GAO-05-805 (Washington, D.C.: Sept. 7, 2005) and *Census Bureau: Important Activities for Improving Management of Key 2010 Decennial Acquisitions Remain to be Done,* GAO-06-444T (Washington, D.C.: Mar. 1, 2006).

[11] Homeland Security Act of 2002, Pub. L. No. 107-296 (Nov. 25, 2002).

[12] The critical infrastructure sectors include agriculture and food; banking and finance; chemical; commercial facilities; commercial nuclear reactors, materials, and waste; communications; critical manufacturing; dams; defense industrial base; drinking water and water treatment systems; emergency services; energy; government facilities; information technology; national monuments and icons; postal and shipping; public health and health care; and transportation systems.

[13] GAO, *Information Technology: Numerous Federal Networks Used to Support Homeland Security Need to Be Better Coordinated with Key State and Local Information-Sharing Initiatives*, GAO-07-455 (Washington, D.C.: April 16, 2007).

[14] GAO, *Critical Infrastructure Protection: Sector Plans and Sector Councils Continue to Evolve,* GAO-07-706R (Washington, D.C.: July 10, 2007).

[15] The department issued a cost-plus-fixed-fee task order under the Enterprise Acquisition Gateway for Leading Edge Solutions (EAGLE). EAGLE is a DHS multiple award indefinite-delivery/indefinite-quantity contract, under which DHS conducted a competition for the HSIN Next Gen task order.

[16] For example, GAO-05-805 and GAO-06-444T.

[17] Carnegie Mellon Software Engineering Institute, *Capability Maturity Model® Integration for Acquisition (CMMI-ACQ),* Version 1.2 (November 2007).

[18] Carnegie Mellon Software Engineering Institute, *Capability Maturity Model® Integration for Acquisition (CMMI-ACQ),* Version 1.2 (November 2007).

INDEX

C

D

H

I

J

O

P

Q

R

S

T

U

V

W